I0055974

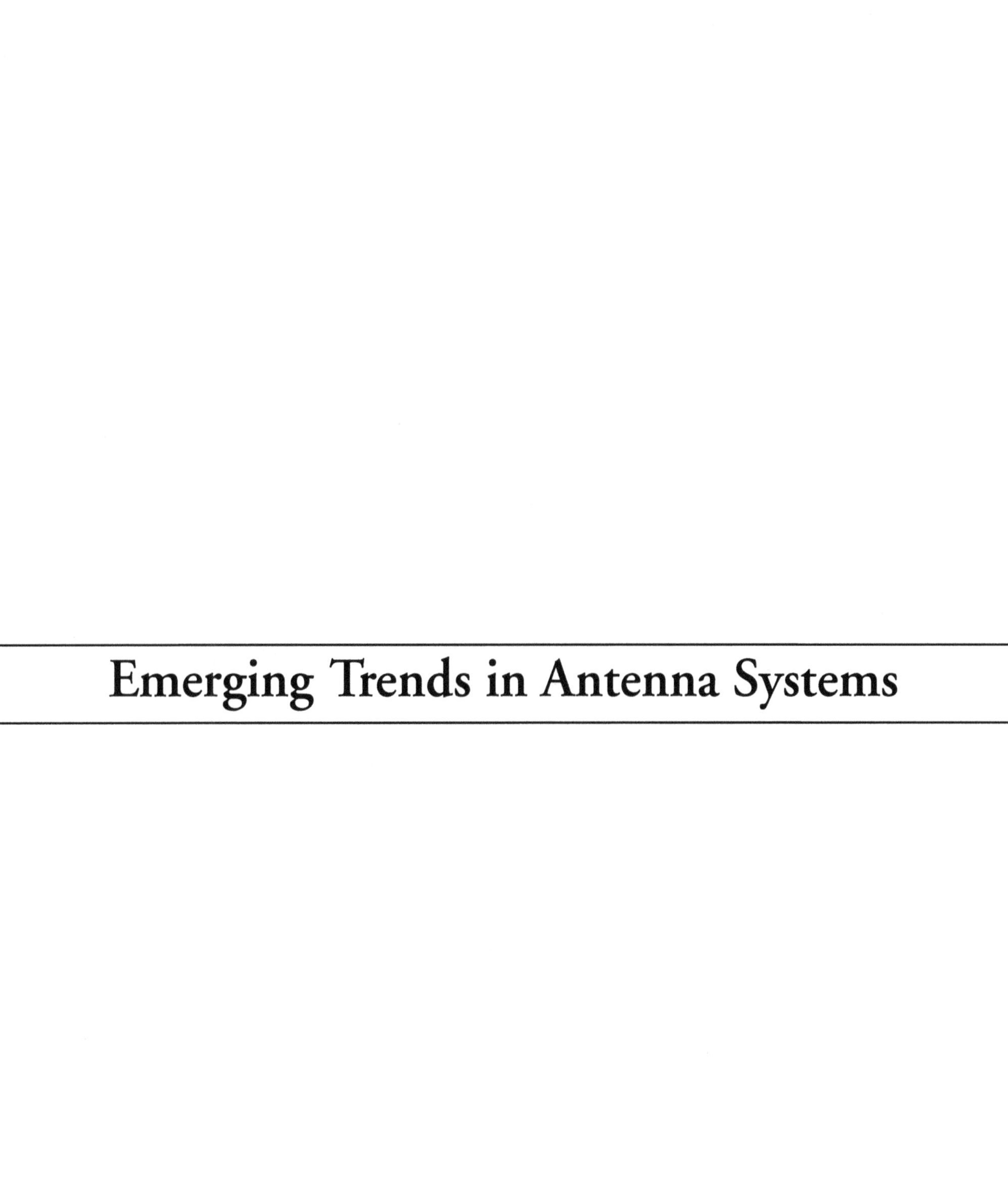

Emerging Trends in Antenna Systems

Emerging Trends in Antenna Systems

Editor: Joe Myers

Published by Murphy & Moore Publishing
1 Rockefeller Plaza,
New York City, NY 10020, USA
www.murphy-moorepublishing.com

Emerging Trends in Antenna Systems
Edited by Joe Myers

© 2022 Murphy & Moore Publishing

International Standard Book Number: 978-1-63987-189-6 (Hardback)

This book contains information obtained from authentic and highly regarded sources. Copyright for all individual chapters remain with the respective authors as indicated. All chapters are published with permission under the Creative Commons Attribution License or equivalent. A wide variety of references are listed. Permission and sources are indicated; for detailed attributions, please refer to the permissions page and list of contributors. Reasonable efforts have been made to publish reliable data and information, but the authors, editors and publisher cannot assume any responsibility for the validity of all materials or the consequences of their use.

Trademark Notice: Registered trademark of products or corporate names are used only for explanation and identification without intent to infringe.

Cataloging-in-Publication Data

Emerging trends in antenna systems / edited by Joe Myers.
p. cm.
Includes bibliographical references and index.
ISBN 978-1-63987-189-6
1. Antennas (Electronics). 2. Electronic apparatus and appliances. I. Myers, Joe.
TK7871.6 .E44 2022
621.382 4--dc23

Contents

Preface VII

Chapter 1 **Planar Antennas with Enhanced Bandwidth and Radiation Characteristics** 1
Mohammad Alibakhshikenari, Mohammad Naser-Moghadasi, Ramazan Ali Sadeghzadeh, Bal Singh Virdee and Ernesto Limiti

Chapter 2 **A Circularly Polarized Spiral/Loop Antenna and its Simple Feeding Mechanism** 15
Mayumi Matsunaga

Chapter 3 **Omnidirectional Circularly Polarized Antenna with High Gain in Wide Bandwidth** 30
Bin Zhou, Junping Geng, Xianling Liang, Ronghong Jin and Guanshen Chenhu

Chapter 4 **Investigating EM Dipole Radiating Element for Dual-Polarized Phased Array Weather Radars** 50
Ridhwan Khalid Mirza, Yan (Rockee) Zhang, Dusan Zrnic and Richard Doviak

Chapter 5 **Application of Composite Right/Left- Handed Metamaterials in Leaky Wave Antennas** 72
Keyhan Hosseini and Zahra Atlasbaf

Chapter 6 **Micro Switch Design and its Optimization Using Pattern Search Algorithm for Application in Reconfigurable Antenna** 96
Paras Chawla and Rohit Anand

Chapter 7 **Recent Computer-Aided Design Techniques for Rectangular Microstrip Antenna** 118
Sudipta Chattopadhyay and Subhradeep Chakraborty

Chapter 8 **Metamaterial Antennas for Wireless Communications Transceivers** 145
Mohammad Alibakhshikenari, Mohammad Naser-Moghadasi, Ramazan Ali Sadeghzadeh, Bal Singh Virdee and Ernesto Limiti

Chapter 9 **Compact Antenna with Enhanced Performances Using Artificial Meta-Surfaces** 161
Tong Cai, He-Xiu Xu, Guang-Ming Wang and Jian-Gang Liang

Permissions

List of Contributors

Index

Preface

The interface between radio waves that propagate through space and electric currents moving in metal conductors is termed as an antenna. It is used with a transmitter or receiver. In transmission process, electric current is supplied to the antenna's terminals through a radio transmitter, and the antenna radiates energy from the current in the form of electromagnetic waves or radio waves. During the process of reception, an antenna catches some of the power of a radio wave so that it can produce electric current at its terminals, which is conveyed to a receiver to be amplified. Some of the important characteristics of antennas are bandwidth, gain, radiation pattern efficiency and polarization. This book elucidates the concepts and innovative models around prospective developments with respect to antenna systems. It explores all the important aspects of antenna systems in the present day scenario. As this field is emerging at a rapid pace, the contents of this book will help the readers understand the modern concepts and applications of the subject.

This book is a comprehensive compilation of works of different researchers from varied parts of the world. It includes valuable experiences of the researchers with the sole objective of providing the readers (learners) with a proper knowledge of the concerned field. This book will be beneficial in evoking inspiration and enhancing the knowledge of the interested readers.

In the end, I would like to extend my heartiest thanks to the authors who worked with great determination on their chapters. I also appreciate the publisher's support in the course of the book. I would also like to deeply acknowledge my family who stood by me as a source of inspiration during the project.

Editor

1

Planar Antennas with Enhanced Bandwidth and Radiation Characteristics

Mohammad Alibakhshikenari,
Mohammad Naser-Moghadasi,
Ramazan Ali Sadeghzadeh, Bal Singh Virdee and
Ernesto Limiti

Abstract

Wireless companies want next-generation gadgets to download at rates of gigabits per second. This is because there is an exponential growth in mobile traffic, however, existing digital networks and devices will not be efficient enough to handle this much growth. In order to realize this requirement, the next generation of wireless communication devices will need to operate over a much larger frequency bandwidth. In this chapter, novel wideband and ultra-wideband (UWB) antennas that are based on loading the background plane of a monopole radiator with concentric split-ring resonators are presented. It is shown that this modification improves the fractional bandwidth of the antenna from 41 to 87%; in particular, the operational bandwidth of the proposed antennas is double that of a conventional monopole antenna of the same size.

Keywords: split-ring resonator, ultra-wideband antenna, composite right/left-handed metamaterial, miniaturized antennas

1. Introduction

Multiband antennas for the personal wireless communications that facilitate various global communication standards and services have become imperative [1, 2]. The antennas for modern communication devices should be capable of operating at maximum possible frequency bands to serve multiple cellular and noncellular communication applications. Moreover, the antennas for portable handheld wireless communication terminals should have compact, low profile, robust, light weight, easy to manufacture, and flexible [3, 4].

The introduction of metamaterials (MTM) has provided unique and naturally nonexistent properties that have enabled the design of antennas with sizes much smaller than their conventional counterparts [5, 6]. Metamaterial structures can be realized using the resonant and nonresonant approaches [7, 8]. In particular, metamaterial structures can be implemented using the split-ring resonators (SRRs), complementary SRR (CSRRs), transmission-line-based structures, electric-field coupled-LC (ELC), etc. [9]. The properties of metamaterials are widely exploited to realize compact antennas with improved performance [10, 11]. Although resonant structures are inherently narrow band, a number of studies have been conducted to improve the bandwidth of metamaterial antennas [12, 13].

This chapter presents a number of antennas implemented with asymmetrical SRR loading and without SRR loading. The first version of antenna consists of a standard monopole radiator without SRR loading, and the second version comprises two monopoles with SRR loading. In effect, with the SRR loading, the bandwidth of the F-shaped and T-shaped antennas improves considerably. The F-shaped antenna's bandwidth improves from 31.5 to 75.4%, and in the case of the T-shaped antenna, the improvement is from 41.16 to 86.9% for VSWR ≤2. The SRR loading pulls down the resonant frequency so that both antennas cover the required frequency range related to worldwide interoperability for microwave access (WiMAX 3.5/5.5 GHz) and wireless local area network (WLAN 5.2/5.8 GHz) bands.

2. Antenna configurations

Two antennas with differing configurations with and without asymmetrical split-ring loading on the ground plane are shown in **Figures 1** and **2**, respectively. The F-shaped and T-shaped antenna structures are essentially monopole radiators. The second versions of the antennas were loaded with asymmetrical SRR on the ground plane to enhance their radiation characteristics. The ground planes of the antennas are truncated into L-shape with a notch.

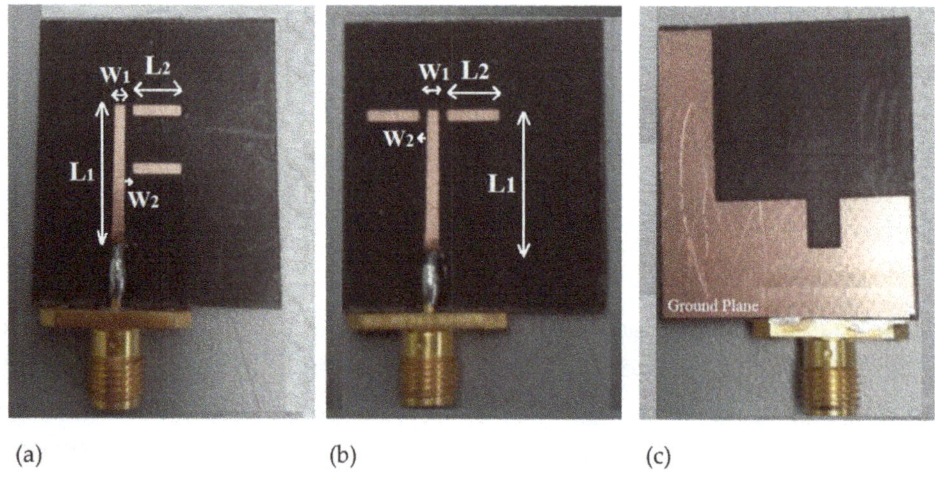

(a) (b) (c)

Figure 1. Fabricated prototypes of the monopole antennas, (a) top view of F-shaped antenna, (b) top view of T-shaped antenna, and (c) back view of both antennas without SRR loading. The antennas are in the *xy* plane.

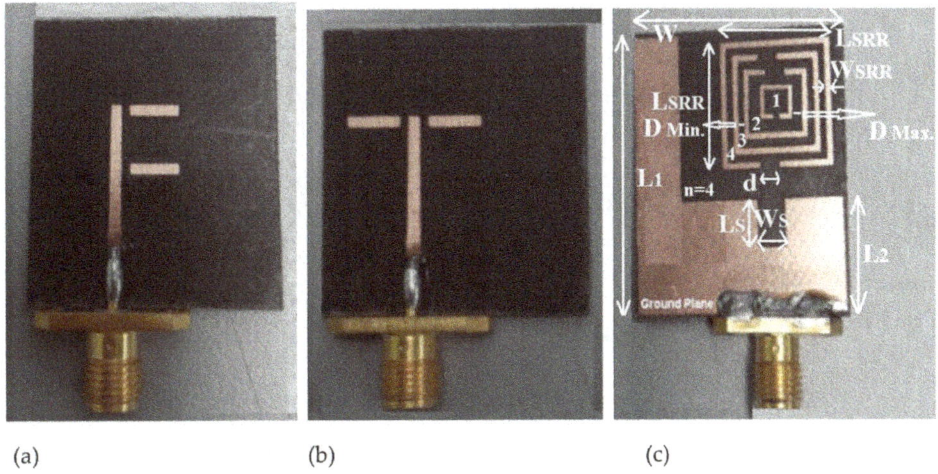

Figure 2. Fabricated prototypes of the antennas with asymmetrical SRR loading, (a) top view of F-shaped antenna, (b) top view of T-shaped antenna, and (c) back view of both antennas with SRR loading.

This configuration improved the antennas impedance matching. To validate the designs, the antenna structures were modeled and analyzed by Ansys 3D full-wave electromagnetic field simulator called high-frequency structure simulator (HFSS™) [14]. The antennas were optimized and fabricated on RT/duroid® 5880 substrate permittivity (ε_r) of 2.2, loss tangent (tanδ) of 0.002, and thickness (h) of 0.8 mm.

The dimensions of the parameters in **Figure 1** are as follows: L_1 = 19 mm, L_2 = 7 mm, W_1 = 1.4 mm, and W_2 = 0.4 mm. Dimensions in **Figure 2c** are as follows: L_{SRR} = 12 mm, W_{SRR} = 0.4 mm, D_{Min} = 0.4 mm, D_{Max} = 0.8 mm, d = 1.5 mm, L_1 = 30 mm, L_2 = 11 mm, W = 22 mm, L_S = 3.5 mm, and W_S = 2.2 mm. Overall size of the antennas are $25 \times 10^{-2}\lambda_o \times 11 \times 10^{-2}\lambda_o \times 2 \times 10^{-2}\lambda_o$ for F-shaped antenna and $25 \times 10^{-2}\lambda_o \times 21 \times 10^{-2}\lambda_o \times 2 \times 10^{-2}\lambda_o$ for T-shaped antenna, where the free space wavelength (λ_o) is 4 GHz. The performance of the antennas was verified using a network analyzer.

Asymmetrical split-ring resonator loading on the F-shaped and T-shaped antennas, shown in **Figure 2**, excites a lower resonance frequency mode that matches with the resonance frequency of monopole antennas, thus increasing the bandwidth of the antennas. As a result, the operational bandwidth of the antennas is extended to cover WiMAX (3.5–5.5 GHz) and WLAN (5.2–5.8 GHz) bands.

2.1. Results and discussion

The performance of the F-shaped and T-shaped antennas is compared with and without asymmetrical SRR loading. The simulated and measured reflection coefficient of the two antennas with and without SRR loading is shown in **Figure 3**. The simulated and measured impedance bandwidth of the F-shaped antenna without SRR loading is 1.62 GHz (3.95–5.57 GHz) and 1.5 GHz (4.0–5.50 GHz), respectively, which corresponds to a fractional bandwidth of 34.0 and 31.5%, respectively. The simulated and measured impedance bandwidth of the T-shaped antenna without SRR loading is 2.08 GHz (3.77–5.85 GHz) and 1.98 GHz (3.82–5.80 GHz), respectively, which corresponds to a fractional bandwidth of 43.24 and 41.16%, respectively. These results also show that there is excellent agreement in the simulated and measured results.

Results in **Figure 3** show that by loading the ground plane with SRR, a lower resonant frequency mode is excited in both antennas that match with the resonance frequency of monopole antennas. This has a consequence of increasing the bandwidth of the antennas. Simulation results using HFSS™ show that with SRR loading, the F-shaped and T-shaped antennas resonate at 3.65, 5.8 and 4.1, 5.9 GHz, respectively, and the measured resonant modes for the F-shaped and T-shaped antennas occur at 3.75, 5.9 GHz and 4, 6 GHz, respectively.

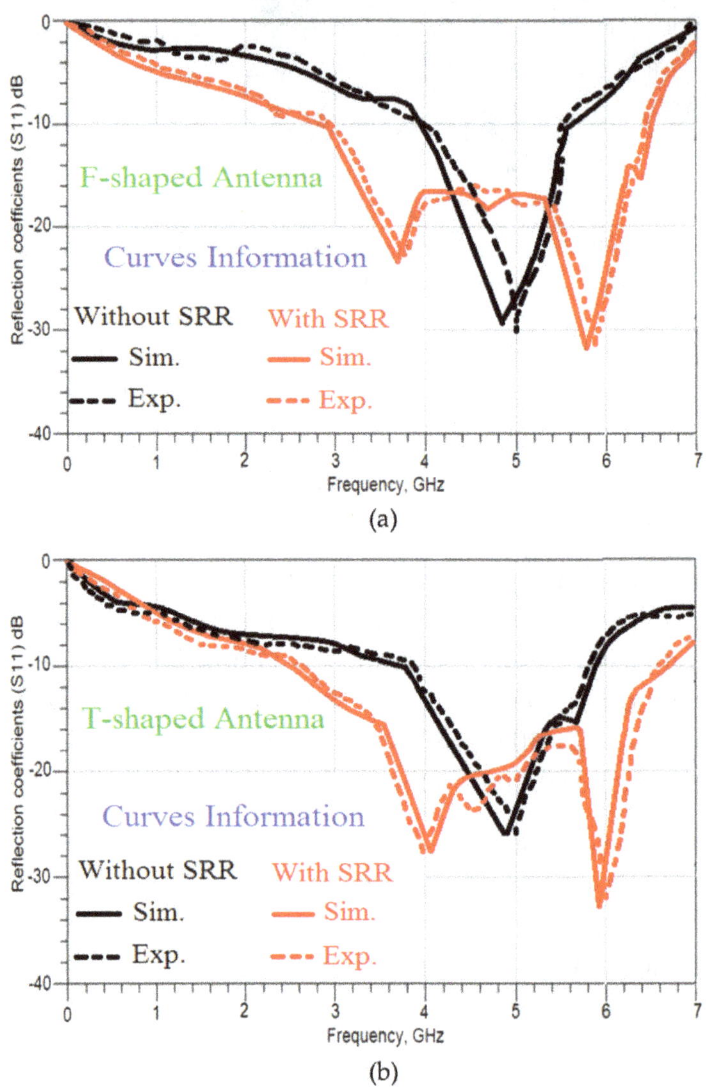

Figure 3. Simulated and measured reflection coefficient of (a) F-shaped and (b) T-shaped antennas with and without SRR loading.

It is evident from **Figure 3a** that the F-shaped antenna loaded with SRR has an impedance bandwidth of 78.27% (2.83–6.47 GHz) using HFSS™. The actual impedance bandwidth measured is 75.40% (2.9–6.41 GHz). Similar performance is obtained with the T-shaped antenna when loaded with SRR, as shown in **Figure 3b**. The simulated and measured impedance bandwidth of the SRR-loaded T-shaped antenna is 4.11 GHz (2.57–6.68 GHz) and 4 GHz (2.6–6.6 GHz), respectively, which corresponds to a fractional bandwidth of 88 and 87%, respectively. The operating frequency range of F-shaped and T-shaped antennas without

SRR using HFSS™ is centered at f_r = 4.80 and 4.85 GHz, respectively; however, the measured operating frequency range is centered at 5 GHz in both cases. The results show that with SRR loading on the ground plane of the antennas can considerably improve the bandwidth of both F-shaped and T-shaped antennas. Consequently, it is found that the SRR loading extends the impedance bandwidth of the monopole antennas enabling coverage across the WiMAX (3.5–5.5 GHz) and WLAN (5.2–5.8 GHz) bands.

For the unloaded antennas, the measured fractional bandwidth is 31.5% for the F-shaped antenna and 41.16% for the T-shaped antenna. With SRR loading, the fractional bandwidth improvement is 75.4% for the F-shaped antenna and 86.9% for the T-shaped antenna. There is clearly an improvement by a factor of 2.39 (F-shaped antenna) and 2.11 (T-shaped antenna).

The measured gain and radiation efficiency of the F-shaped and T-shaped antennas with and without the SRR loading is shown in **Figure 4a** and **b**, respectively, and salient features are given in **Table 1**.

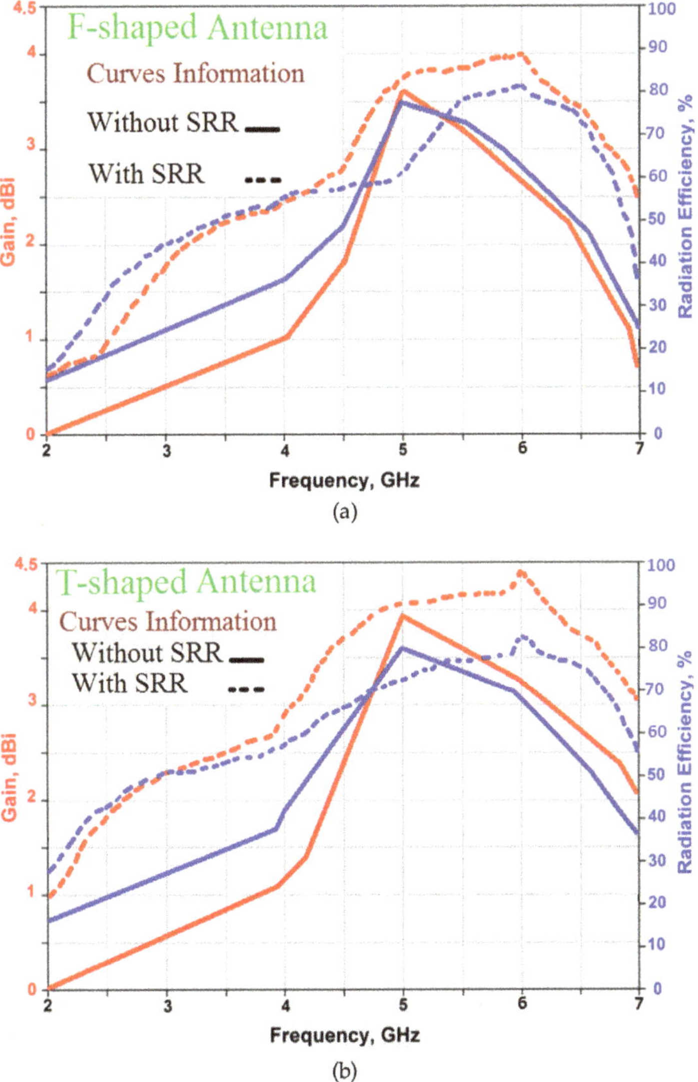

Figure 4. Measured gain and efficiency response of the two antennas with and without SRR loading. (a) F-shaped antenna and (b) T-shaped antenna.

	Without asymmetrical SRR loading		With asymmetrical SRR loading
F-shaped antenna	Gain (dBi)	@ 4, 4.5, **5**, 5.5 GHz: 1, 1.8, **3.6**, and 3.2	@ 2.9, 3.8, 5, **6**, 6.41 GHz: 1.7, 2.4, 3.7, **4**, 3.45
	Efficiency (%)	@ Same Freq.: 36.3, 48.1, **78.5**, 73.8	@ Same Freq.: 43.6, 51.7, 60.9, **81.2**, 76.5
T-shaped antenna	Gain (dBi)	@ 3.82, 4, **5**, 5.8 GHz: 1.15, 1.3, **3.9**, 3.3	@ 2.6, 3.82, 4, 5.8, **6**, 6.6 GHz: 1.8, 2.75, 2.9, 4.25, **4.4**, 3.7
	Efficiency (%)	@ Same Freq.: 36.5, 42.1, **80.2**, 70.4	@ Same Freq.: 45.5, 56.5, 58.9, 79.3, **82.6**, 75.4

Table 1. Radiation characteristics of the two antennas (*maximum values have been set as bold fonts*).

Figure 5a shows the two antennas radiate omni directionally in the E-plane across the entire operating frequency range in both cases of with and without SRR loading. Cross polarization is at least 15 dB below the main radiation. Both antennas radiate bidirectionally in the H-plane across their operating frequency. In this case, the cross polarization is also at least 15 dB below the main radiation. Results show that the gain is much stronger for both antennas with SRR loading at 6 GHz, and at 5 GHz, the gain is more pronounced only for the T-shaped antenna loaded with SRR.

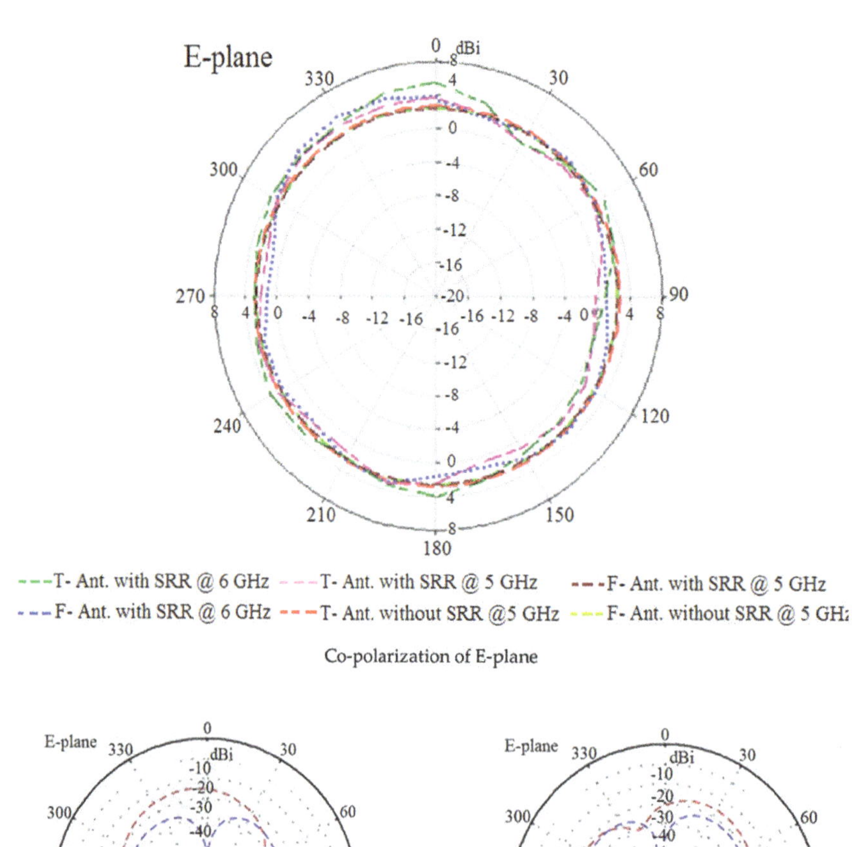

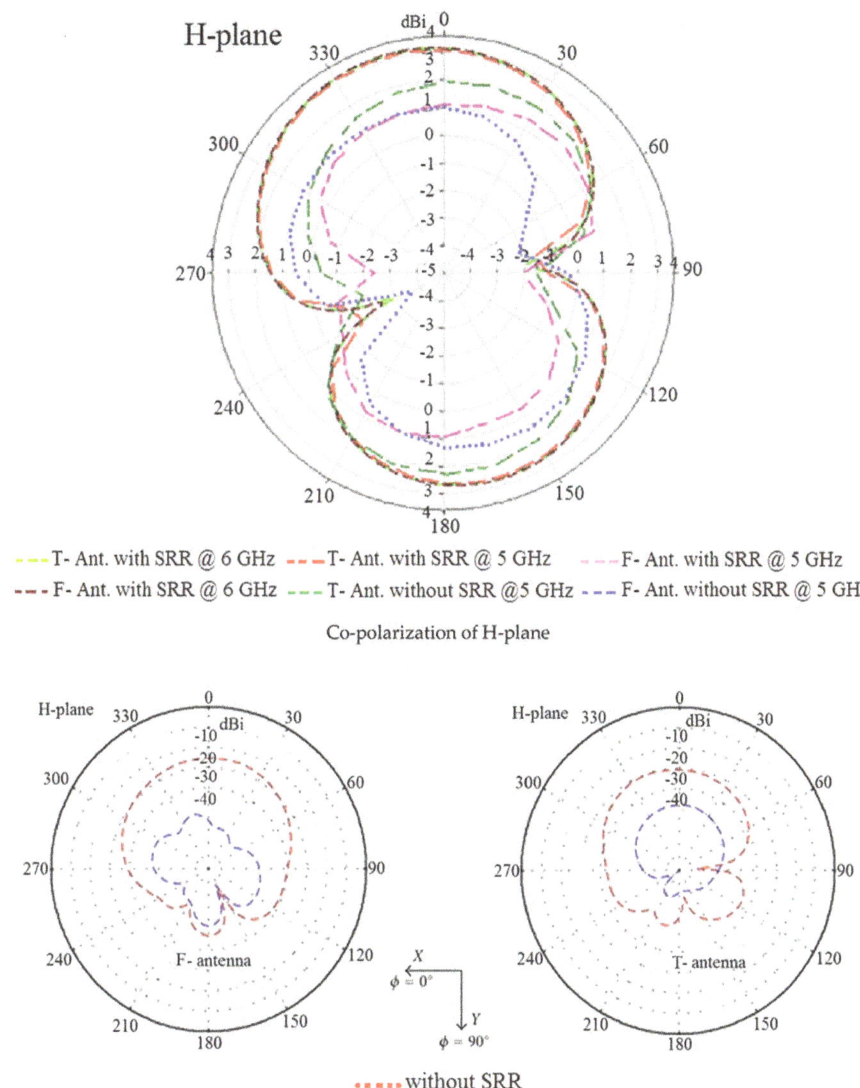

Figure 5. Measured E-plane and H-plane radiation patterns for F-shaped and T-shaped antennas with and without asymmetrical SRR loading at the operating frequencies. (a) Measured E-plane radiation patterns for F-shaped and T-shaped antennas. E-plane corresponds to Phi = 0°. (b) Measured H-plane radiation patterns for F-shaped and T-shaped antennas. H-plane corresponds to Theta = 90°. The slight shift in beam direction in H-plane is due to loading of SRR on the ground plane.

2.2. Effect of asymmetrical SRR loading

The propagation characteristics of the proposed structures are now considered by analyzing their dispersion properties. **Figure 6** shows the simplified discrete element equivalent circuit of the antenna structures. This electrical circuit is valid in the long-wavelength region (i.e., $\beta l \ll 1$). Note, β is the propagation constant for guided waves, and l is the length of the structure.

L (H/m) and C (F/m) represent the inductance and capacitance, respectively. These two reactance components define the inductance associated with the central conductor in the structures and the capacitive coupling between the structures and ground planes. SRR loading is modeled by a parallel resonant circuit constituted from inductance (L_S) and capacitance (C_S). The parallel resonant circuit is electromagnetically coupled to the structure through a mutual inductance (*M*). It can be shown that the dispersion relation is given by:

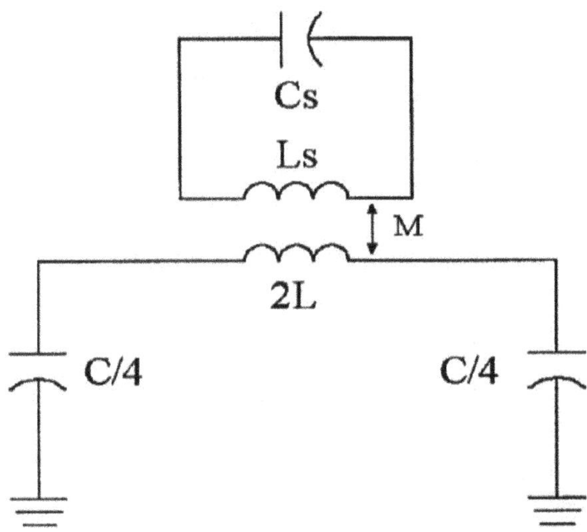

Figure 6. Lumped element equivalent circuit for the antenna structures.

$$\cos(\beta l) = 1 - \frac{LC\omega^2}{2} + \frac{C/C'_S}{4(1-\omega_0^2/\omega^2)} \tag{1}$$

$$C'_S = \frac{L_S}{M^2\omega_0^2} \tag{2}$$

$$L'_S = C_S M^2 \omega_0^2 \tag{3}$$

$$\omega_0^2 = \frac{1}{L_S C_S} = \frac{1}{L'_S C'_S} \tag{4}$$

M can be inferred from the fraction (A_f) of the slot area occupied by the rings according to:

$$M = 2L\,A_f \tag{5}$$

By obtaining the equivalent impedance of the series branch, the circuit can be simplified as shown in **Figure 7**. The dispersion relation for the circuit in **Figure 7** is given by [15]:

$$\cos(\beta l) = 1 - \left(\frac{L\omega - 1/C\omega}{4L/C}\right)\left(2L\omega - \frac{L'_S/C'_S}{L'_S\omega - 1/C'_S\omega}\right) \tag{6}$$

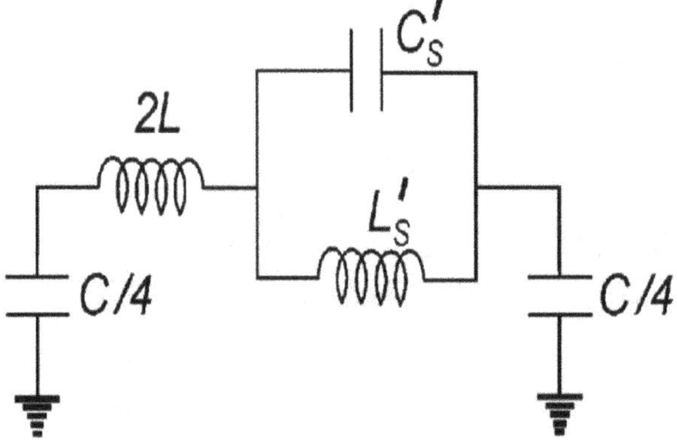

Figure 7. Simplified circuit with the series branch replaced by its equivalent impedance.

The dispersion diagram for the corresponding structures using HFSS™ and Eq. (6) is shown in an ω–β diagram in **Figure 8**. The circuit elements have been calculated for the antenna structures shown in **Figure 2**.

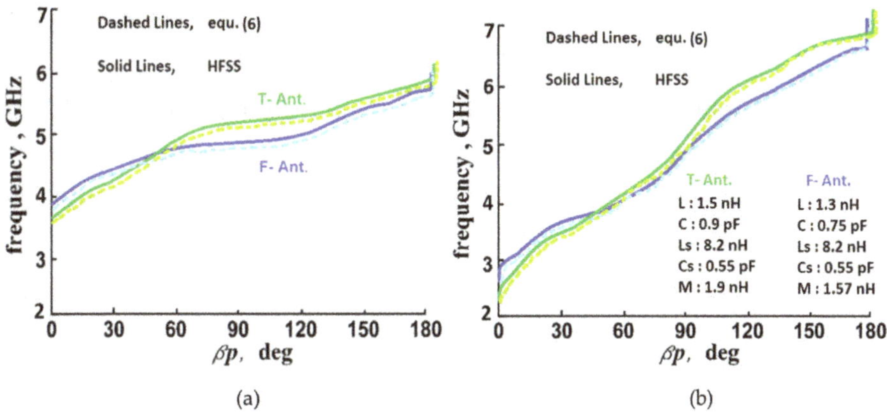

Figure 8. Dispersion diagram of the proposed F-shaped and T-shaped antennas. (Blue lines are for F-shaped antenna, and the green lines are for T-shaped antenna). (a) Without SRR loading and (b) loaded with SRR.

Nicolson-Ross-Weir method was used to obtain the permeability of the SRR [16]. Impedance and permeability response of F-shaped and T-shaped antennas loaded with SRR, shown in **Figure 9**, reveal that the impedance and permeability of the SRR in both antennas are positive and stable over the operating bands of antennas.

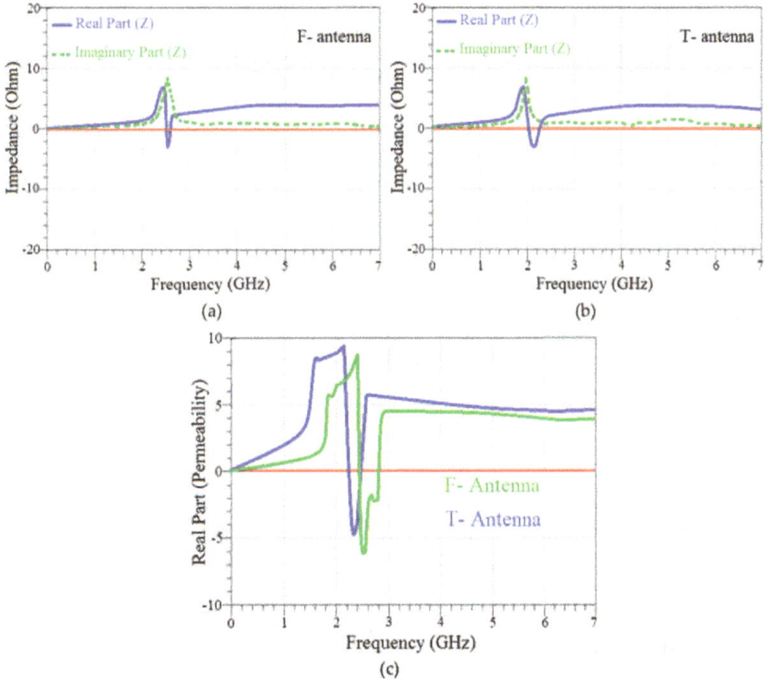

Figure 9. Impedance and permeability response for the two antennas loaded with SRR. (a) Impedance of the SRR for F-shaped antenna, (b) impedance of the SRR for T-shaped antenna and (c) permeability of the SRR for both antennas.

To achieve the desired performance from the antennas, it was necessary to optimize the dimensions of the asymmetrical SRR loading using Ansys HFSS™ EM Simulator. **Figures 10** and **11** show the effect of the SRR dimensions on the reflection coefficient response of the two antennas as a function of length (L_{SRR}), width (W_{SRR}), and gap between the SRR rings (D_{SRR}). Results show that when the length (L_{SRR}), width (W_{SRR}), and gap (D_{SRR}) of SRR loading are increased, the bandwidth of the antennas is marginally improved.

Effect on the antenna's gain and radiation efficiency as a function of the SRR dimensions is shown in **Figure 12**. These results clearly show the dimensions of the SRR greatly influences the antenna's the gain and efficiency properties over the entire operating frequency range of the antenna.

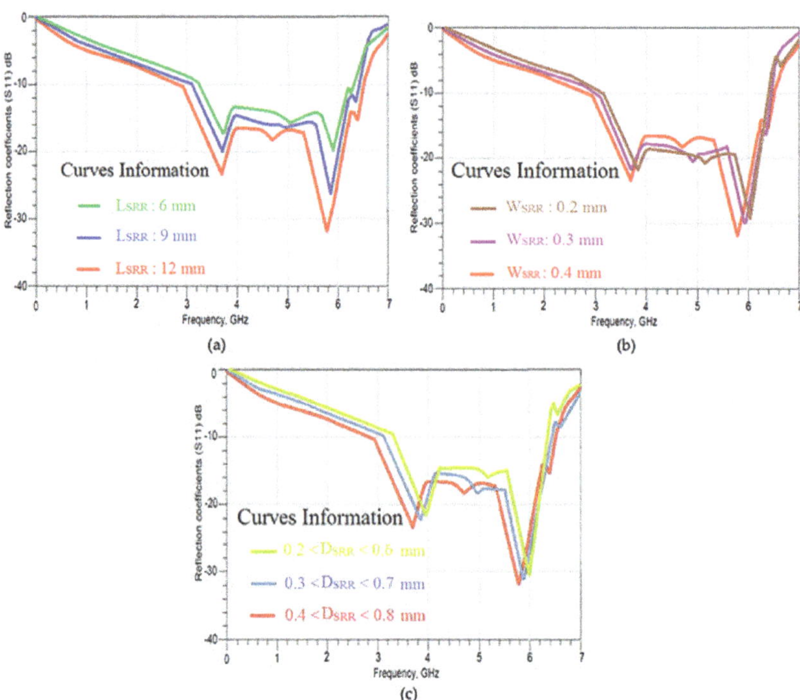

Figure 10. Parametric study on SRR dimensions for the F-shaped antenna. (a) Effect of SRR length (L_S) (all other parameters remain constant). (b) Effect of SRR width (W_S) (all other parameters remain constant). (c) Effect of gap between the rings of the SRR (D_{SRR}) (all other parameters remain constant).

3. Conclusion

To summarize, this chapter presented two monopole antenna configurations loaded with and without split-ring resonators (SRR). The results presented show that SRR loading improves the antenna's performance in terms of bandwidth and radiation characteristics compared to conventional unloaded monopole antennas. This is because SRR-loaded antennas excite a lower resonance frequency mode that matches with the resonance frequency of the monopoles, thereby extending the overall impedance bandwidths. The metamaterial antenna presented is implemented with CRLH-TL unit-cells comprising "LT-shaped" slots in a radiating patch with an inductive spiral ground through a via hole. The SRR-loaded antennas operate over a frequency range that covers WiMAX (3.5/5.5 GHz) and WLAN (5.2/5.8 GHz) bands. The antennas are relatively small and have a size of 0.29 $\lambda_0 \times 0.21 \lambda_0$ where λ_0 is 2.9 GHz. The antenna has a maximum gain and radiation efficiency of 4.4 dBi and 82.6%, respectively, at 6 GHz. The metamaterial antenna covers a frequency bandwidth extending from 0.5 to 11.3 GHz for VSWR < 2 and has a size of 22.6 × 5.8 × 0.8 mm^3. It has a maximum gain and radiation efficiency of 6.5 dBi and 88%, respectively, at 8 GHz.

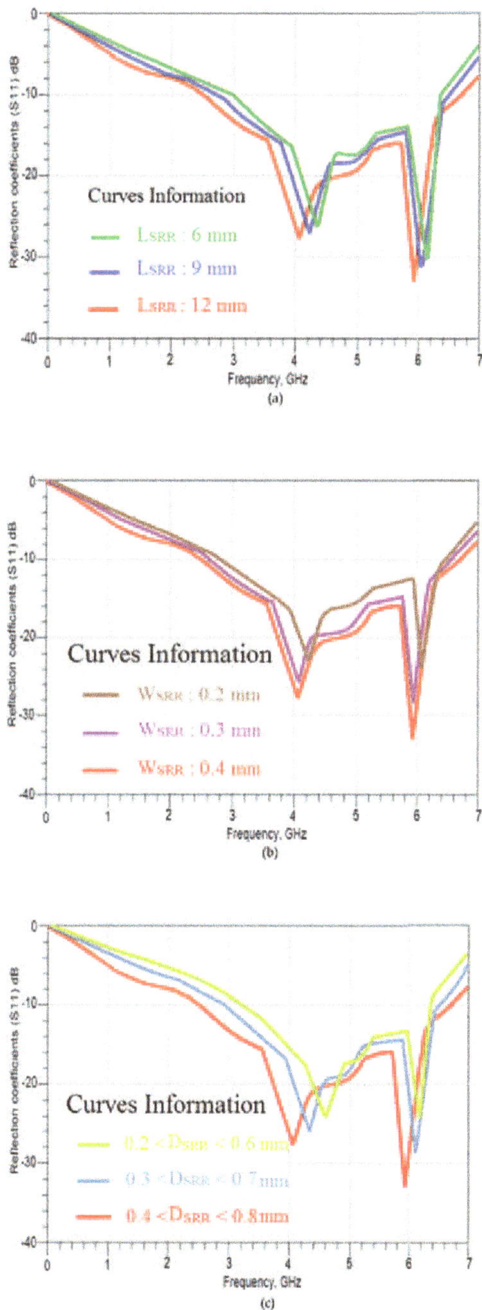

Figure 11. Parametric study on SRR dimensions for the T-shaped antenna. (a) Effect of SRR length (L_S) on the reflection coefficient response (all other parameters remain constant). (b) Effect of SRR width (W_S) on the reflection coefficient response (all other parameters remain constant). (c) Effect of varying the gap between the rings of the SRR (D_{SRR}) on reflection coefficient response (all other parameters remain constant). The number of SRR pairs used is four.

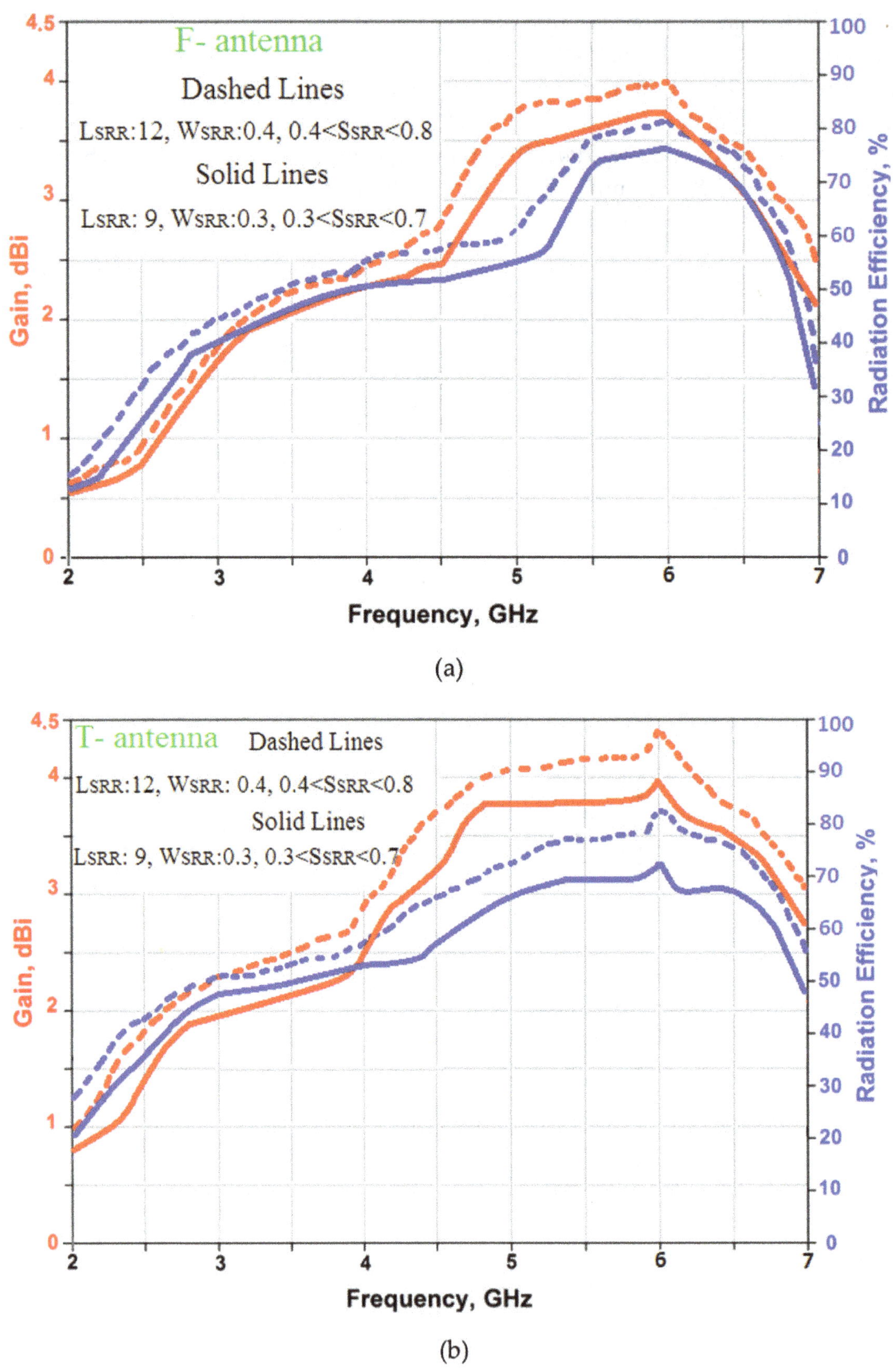

Figure 12. Effect on the radiation characteristics of the two antennas by the dimensions of SRR loading. (a) F-shaped antenna and (b) T-shaped antenna.

Acknowledgements

The Authors would like to give their special thanks to faculty of microelectronics for the financial support.

Author details

Mohammad Alibakhshikenari[1*], Mohammad Naser-Moghadasi[2], Ramazan Ali Sadeghzadeh[3], Bal Singh Virdee[4] and Ernesto Limiti[1]

*Address all correspondence to: Alibakhshikenari@ing.uniroma2.it

1 Department of Electronic Engineering, University of Rome Tor Vergata, Rome, Italy

2 Faculty of Engineering, Science and Research Branch, Islamic Azad University, Tehran, Iran

3 Faculty of Electrical Engineering, K. N. Toosi University of Technology, Tehran, Iran

4 Center for Communications Technology, London Metropolitan University, London, UK

References

[1] Y. Jee, Y.M. Seo, "Triple-band CPW-fed compact monopole antennas for GSM/PCS/DCS/WCDMA applications," *Electron Lett*, 2009; 45(9): 446–448. doi:10.1049/el.2009.3383

[2] C.T. Lee, K.L. Wong, "Uniplanar printed coupled-fed PIFA with a band-notching slit for WLAN/WiMAX operation in the laptop computer," *IEEE Trans Ant Propag*, 2009; 57(4): 1252–1258. doi:10.1109/TAP.2009.2015843

[3] L. Jofre, B.A. Cetiner, F. De Flaviis, "Miniature multi-element antenna for wireless communications," *IEEE Trans Ant Propag*, 2002; 50(5): 658–669. doi:10.1109/TAP.2002.1011232

[4] M. Alibakhshi-Kenari, "Printed planar patch antennas based on metamaterial,"*Int J Electron Lett*, 2014; 37–42. doi:10.1080/21681724.2013.874042

[5] A. Lai, C. Caloz, T. Itoh, "Composite right/left-handed transmission line metamaterials", *IEEE Microw Mag*, 2004; 5(3): 34–50. doi:10.1109/MMW.2004.1337766

[6] C.J. Lee, K.M.K.H. Leong, T. Itoh, "Composite right/left-handed transmission line based compact resonant antennas for RF module integration," *IEEE Trans Ant Propag*, 2006; 54(8): 2283–2291. doi:10.1109/TAP.2006.879199

[7] Christophe Caloz, Tatsuo Itoh, "*Electromagnetic Metamaterials: Transmission Line Theory and Microwave Applications*", ISBN: 978-0-471-66985-2, 376 pages, December 2005, Wiley-IEEE Press.

[8] C. Caloz, T. Itoh, A. Rennings, "CRLH metamaterial leaky-wave and resonant antennas," *IEEE Ant Propag Mag*, 2008; 50(5): 25–39. doi:10.1109/MAP.2008.4674709

[9] A. Alu, F. Bilotti, N. Engheta, L. Vegni, "Sub-wavelength, compact, resonant patch antennas loaded with metamaterials," *IEEE Trans Ant Propag*, 2007; 55(1): 13–25. doi:10.1109/TAP.2006.888401

[10] C. Caloz, T. Itoh, "Novel microwave devices and structures based on the transmission line approach of meta-materials", *Microwave Symposium Digest, 2003 IEEE MTT-S International*, 8-13 June 2003, Pennsylvania Convention Center Philadelphia, Pennsylvania.

[11] M. Schussler, J. Freese, R. Jakoby, "Design of compact planar antennas using LH-transmission lines", Microwave Symposium Digest, 2004 IEEE MTT-S International, 6-11 June 2004.

[12] Mohammad Alibakhshikenari, Mohammad Naser-Moghadasi, R.A. Sadeghzadeh, Bal S. Virdee, Ernesto Limiti, "Traveling-wave antenna based on metamaterial transmission line structure for use in multiple wireless communication applications", AEU - International Journal of Electronics and Communications, Volume 70, Issue 12, December 2016, Pages 1645–1650.

[13] Mohammad Alibakhshikenari, Mohammad Naser-Moghadasi, R.A. Sadeghzadeh, Bal S. Virdee, Ernesto Limiti, "Miniature CRLH-based ultra wideband antenna with gain enhancement for wireless communication applications", ICT Express, Volume 2, Issue 2, June 2016, Pages 75–79.

[14] Ansoft HFSS, www.ansoft.com/products/hf/hfss.

[15] David M. Pozar, "Microwave Engineering, 4th Edition", December 2011, Wiley.

[16] R.W. Ziolkowski, "Design, fabrication, and testing of double negative metamaterials," *IEEE Trans Ant Propag*, 2003; 51: 1516–1529. doi:10.1109/TAP.2003.813622

2

A Circularly Polarized Spiral/Loop Antenna and its Simple Feeding Mechanism

Mayumi Matsunaga

Abstract

In this chapter, a simple spiral/loop antenna radiating circular polarization is introduced. Circularly polarized antennas are complex structures in general because they are constituted by two or more antennas in multilayer structures or employ phase-shifting circuits. For this reason, the circularly polarized antennas are too complex to be applied to mobile communication devices such as radiofrequency identifier (RFID) and global positioning system (GPS) handy terminals. Simpler and easier circularly polarized antennas are necessary for these devices. The presented circularly polarized antenna is so simple that it is printable, and it has only one port that can be fed through a coaxial cable directly. In the first section, the necessity of the circularly polarized antennas for modern antenna systems is explained. Then, the historical conventional antennas are introduced by referring to important publications. In the second section, the novel simple circularly polarized antenna invented by the author will be presented. The basic structure of the presented antenna and its principle will be explained. In the third section, a feeding mechanism to feed the presented antenna through a coaxial cable will be presented. In conclusion, detailed characteristics of the presented antennas will be summarized.

Keywords: circularly polarized antennas, spiral antennas, loop antennas, planar antennas, wire-printed antennas, coaxial cable feedings, RFID, GNSS

1. Introduction

The popularity of circularly polarized (CP) antennas is increasing because the use of CP waves is thought of as a good way to eliminate the null angle of radiofrequency (RF) transmission signals. In particular, the necessity of small and simple CP antennas is increasing by the year as mobile communication devices are being equipped with radiofrequency identifier (RFID) systems and global navigation satellite systems (GNSSs). The recent technological evolution, such as wireless power transmission and wireless links between electric appliances, equips us with the motivation to develop compact and high-performance CP antennas.

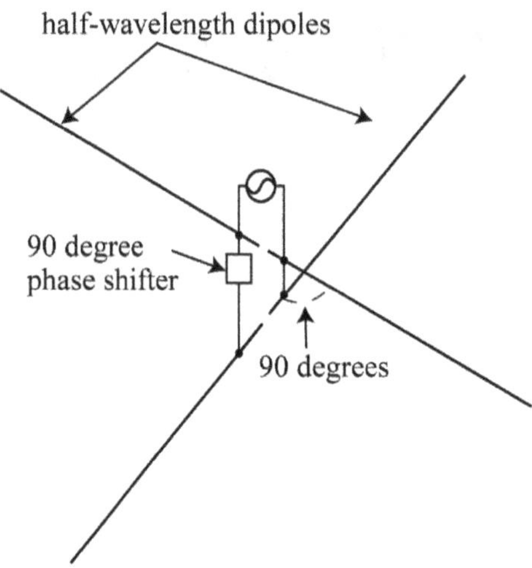

Figure 1. The turnstile antenna invented by Brown, which is recently called the crossed dipole.

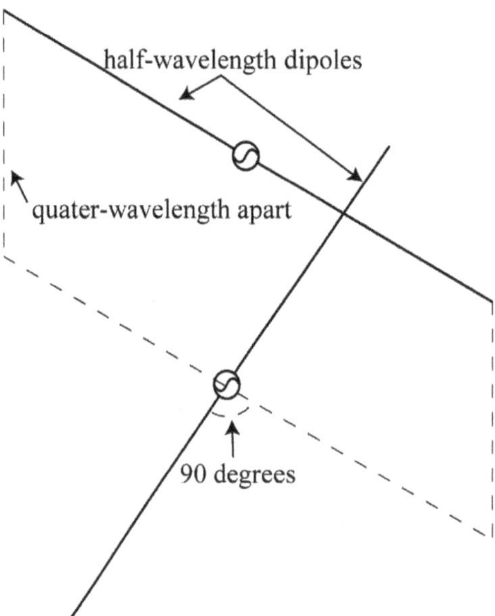

Figure 2. The crossed dipole being spaced by a quarter wavelength instead of using a phase shifter.

The crossed-dipole antenna is the most well-known circularly polarized antenna, which was invented in 1936 by Brown [1–3] as "the turnstile antenna." Brown's idea is that two half wavelength dipoles crossed and fed with 90° phase differences, as shown in **Figure 1**, are the simplest way to make good circularly polarized (CP) waves. Recently, we call this turnstile antenna as the "crossed dipole." In addition, being spaced by a quarter wavelength as shown in **Figure 2**, the crossed-dipole antenna no longer needs to be fed with a 90° phase-shifting circuit. Although the crossed-dipole antenna has become a basic structure of many CP antennas, its beamwidths of CP waves are generally narrow.

Radiating CP waves with wide beamwidths is also a necessary requirement for CP antennas. The Lindenblad antenna [4] is famous as an omnidirectional CP antenna, which is constituted by radially placing two or more sets of crossed dipoles tilted slightly from the transmitting direction, as shown in **Figures 3** and **4**, for omnidirectional CP radiation. In addition, constituted by folded antennas instead of dipoles [5], as shown in **Figure 5**, the Lindenblad antenna radiates with higher gain because its impedance becomes higher. Although the crossed-dipole antenna and the Lindenblad antenna are basic arrangements of CP antennas, it is difficult to make them so compact that they can be installed inside mobile communication devices.

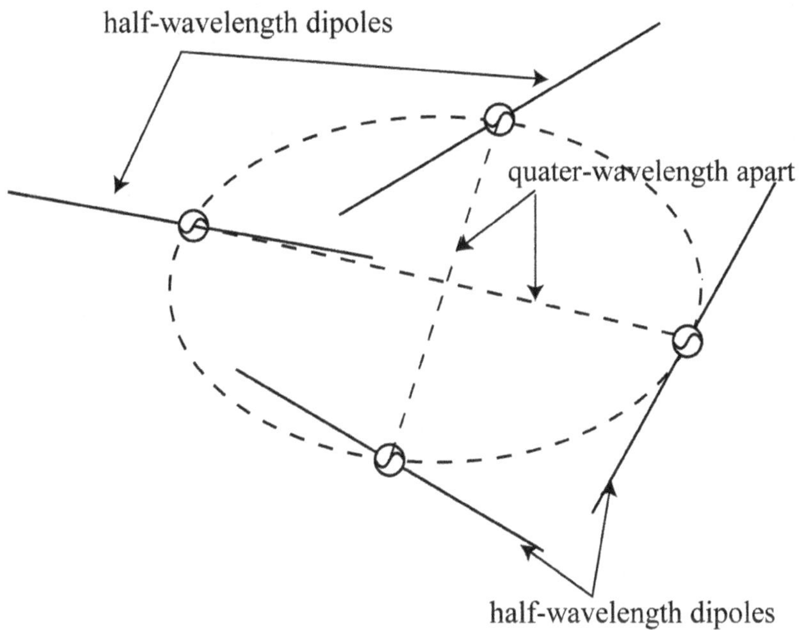

Figure 3. An outline structure of the Lindenblad antenna.

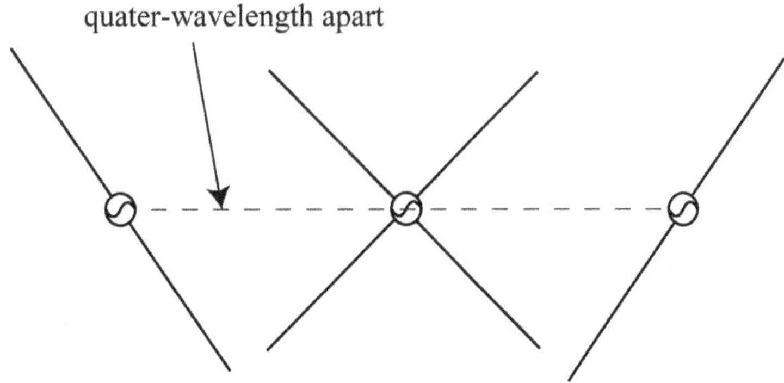

Figure 4. The side view of the Lindenblad antenna shown in **Figure 3**.

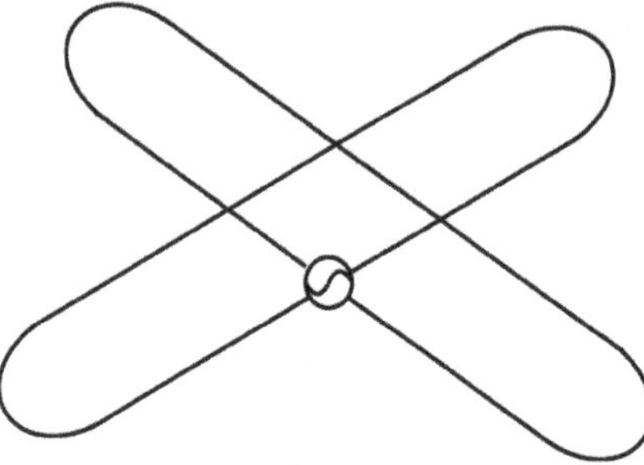

Figure 5. The crossed dipoles or the Lindenblad antenna which is constituted by folded antennas instead of dipoles.

Microstrip patch antennas are a good structure to make thin antennas. Many structures of microstrip patch antennas radiating CP waves were proposed [6]. However, their impedance bandwidths and beamwidths are narrow. Moreover, they need wide enough ground planes. For these reasons, microstrip patch antennas are not suitable for making a printable planar antenna on a one-layer film, which is considered a desirable structure as a built-in antenna for mobile communication devices. Therefore, a new idea for antenna structures is needed for making a compact printable omnidirectional CP antenna.

Based on the previous discussions, the author chose loop antennas as a basic structure for developing a new CP antenna because loop antennas are generally thought of as being thin and capable of being miniaturized easily, having wide impedance bandwidths, wide beamwidths, and high impedance. First, a loop antenna was arranged like a cross shape for radiating CP waves. Second, the cross-shaped loop antenna (CSA) was fed by a dipole antenna for achieving wide beamwidths, multipolarization, and stable feedings. In the next section, the development procedure and principle of these new antennas are explained.

2. Cross-shaped loop antenna

The basic structure of the CP antenna invented by the author is introduced. As mentioned before, the author invented a CP antenna based on loop antennas [7]. The antenna is constituted by a loop antenna arranged like a cross shape; therefore, the antenna is called the "cross-shaped loop antenna (CSA)". Its outline structure is shown in **Figure 6**. The principle of this CSA is based on the crossed-dipole antenna. This means that the antenna is constituted by two crossed elements being fed with a 90° phase difference. Although making a cross-shaped element is easy, feeding it with different phases is a difficult problem.

First, let us think about CSA's structure. As already mentioned, CSA is constituted by arranging a single-turned loop like the cross shape. This means that CSA is completed by turning an electric wire along the outer side of the cross. As a result, CSA is formed by a horizontal element and a vertical element like the crossed-dipole antenna. There are two main differences between CSA and the crossed-dipole antenna: (1) folded antennas are used instead of dipole antennas and (2) CSA is made by connecting folded antennas in series, while the crossed-dipole is made by connecting dipole antennas in parallel.

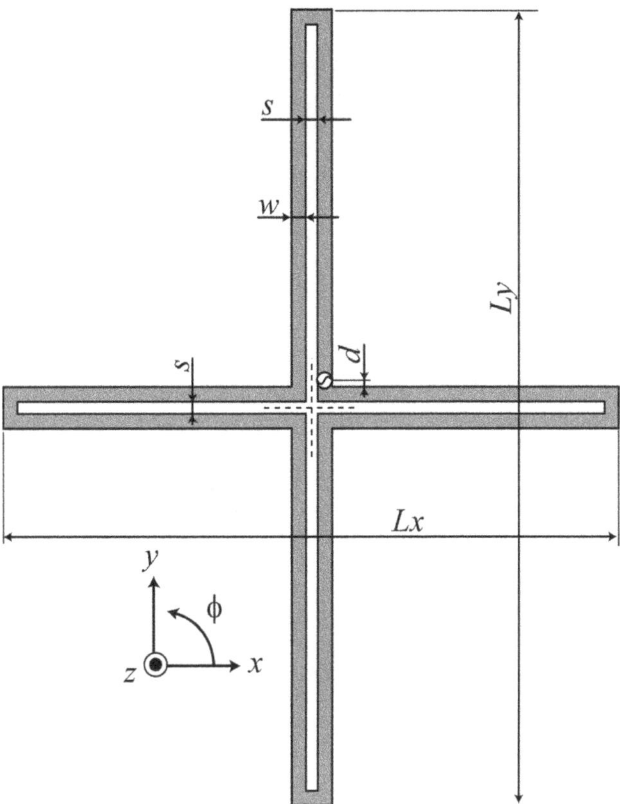

Figure 6. The outline structure of CSA.

Second, let us think about the feeding port of the CSA. To make a compact printable antenna, the horizontal element needs to share its port and plane with the vertical element. The author thought that Bolster's idea [8] would solve this problem. Bolster said that a crossed-dipole antenna can be constituted on the same plane that shares the same port, if the admittance of the horizontal element and that of the vertical element is set to the same conductance, and the arguments have a 90° difference. From this, the author had the idea that CSA could be considered because two-folded antennas are connected in a series as sharing one feeding port, if the feeding port was put in the joint between the upper element and the right element. **Figure 7** shows the equivalent circuit of CSA.

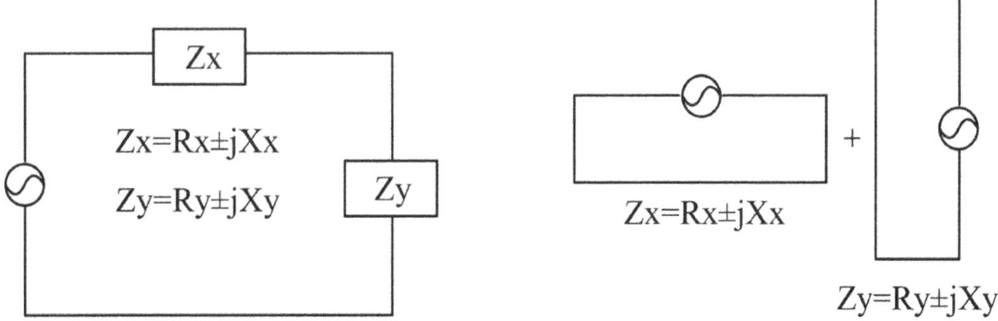

Figure 7. The equivalent circuit of CSA.

According to Bolster's idea, CSA will radiate good CP waves, if the impedances of the horizontal element and vertical element are set as to be close and the arguments are set to a 180° difference. Then, the length of the horizontal folded antenna and that of vertical folded antenna are chosen by **Figure 8**, which shows moduli and arguments of a folded antenna's impedance with various lengths *L*. For example, the horizontal length should be *Lx* marked in **Figure 8**, if the vertical length is chosen as *Ly*. Their impedance moduli $|Zx|$ and $|Zy|$ are close, and impedance arguments $\arg|Zx|$ and $\arg|Zy|$ are in almost 180° difference.

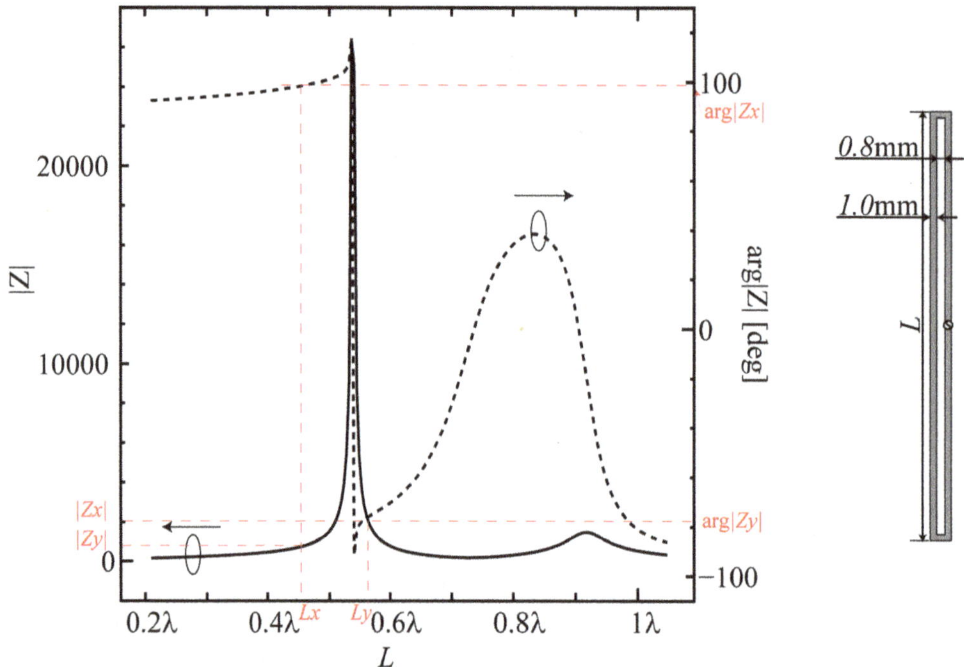

Figure 8. Moduli and arguments of a folded antenna's impedance with various lengths *L*.

Note that the wavelength λ is defined as the following equation (1):

$$\lambda = \frac{c}{f}\sqrt{\varepsilon_r}. \tag{1}$$

f is a frequency, *c* is the light velocity, and ε_r is a relative permittivity of a dielectric substrate on which antennas are constituted. The results in **Figure 8** were calculated when the thickness and relative permittivity of the dielectric substrate were defined as 1.6 mm and 4, respectively. All simulation results shown in the chapter are calculated by Sonnet 16.52.

The characteristics of CSA whose measurements are set based on this procedure are as follows: **Figure 9(a)** shows S_{11} characteristics and **Figure 9(b)** shows radiation patterns. Note that the detailed measurements of CSA are shown in **Table 1**, and λ is the wavelength of the center frequency f_0. CSA radiates right-handed CP waves around the frequencies in which S_{11} is lower than −10 dB.

L_x	L_y	s	w	d
0.448 λ	0.561λ	0.8 mm	1.0 mm	0.2 mm

Table 1. The detailed measurements of the CSA, when $\varepsilon_r = 4.0$ and thickness $t = 1.6$ mm of the dielectric substrate.

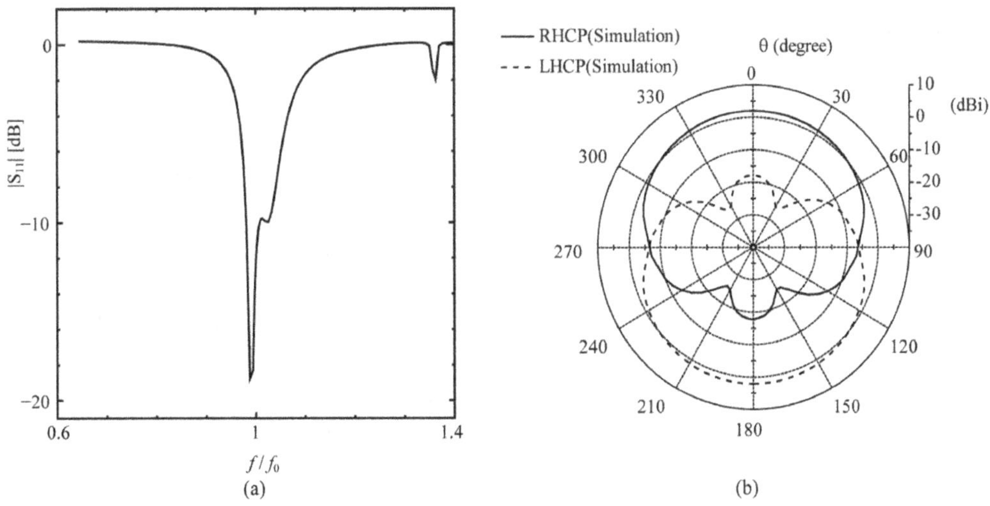

Figure 9. Characteristics of CSA whose measurements are set as shown in **Table 1: (a)** S_{11} characteristics, and **(b)** radiation patterns at $f = f_0$ and $\phi = 0°$.

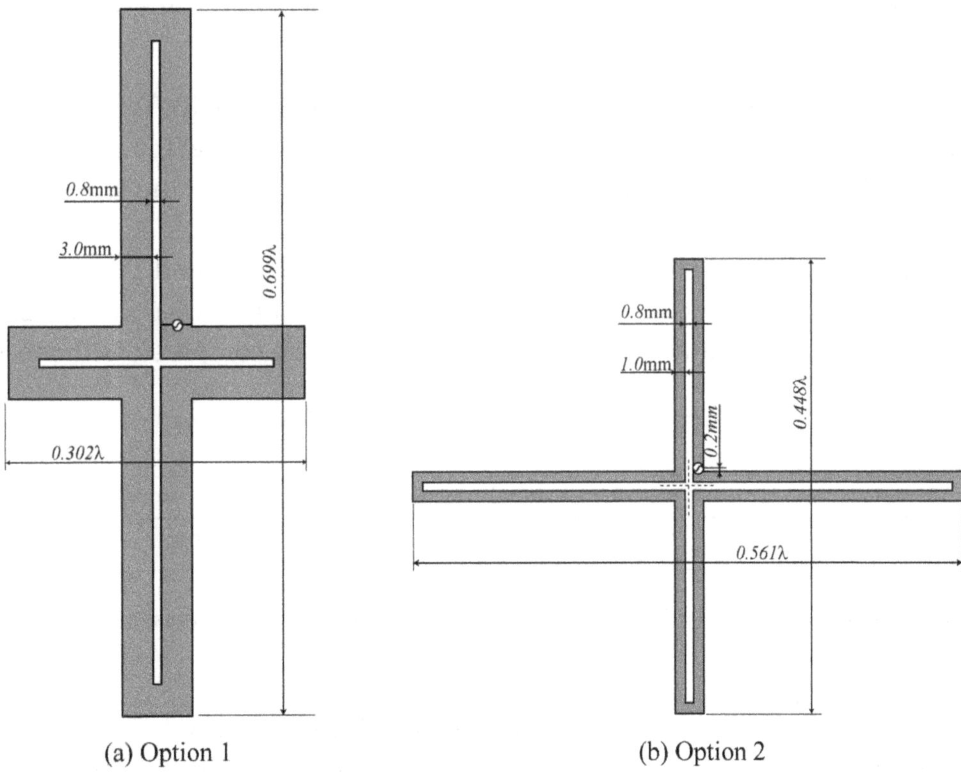

Figure 10. Two optional structures of CSA.

In fact, there are some more possible combinations of the horizontal length and vertical length. **Figure 10** shows two additional combinations of those lengths, and **Figure 11** shows characteristics of these CSAs. This means that you can choose the CSA from two optional structures depending on the space: a rectangle or a square space, assigned to the CSA. Moreover, if the horizontal length is shorter than the vertical length, the CSA will radiate right-hand CP (RHCP) waves. On the other hand, if the horizontal length is longer than the vertical length, CSA will radiate left-hand CP (LHCP) waves.

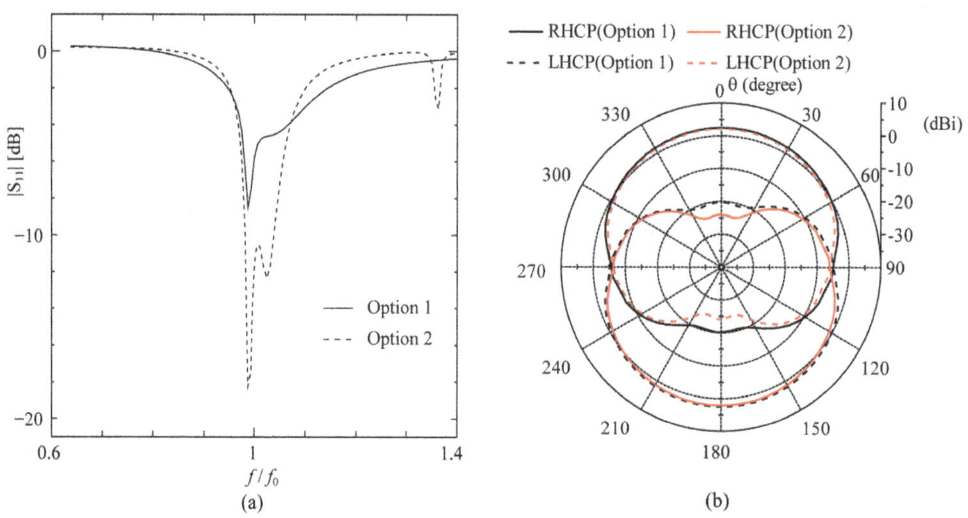

Figure 11. Characteristics of the two optional structures of CSA as shown in **Figure 10: (a)** S_{11} characteristics, and (**b**) radiation patterns at $f = f_0$ and $\phi = 0°$.

Let us arrange the CSA prototype using a printed circuit board (PCB). Wavelengths shorten when the CSA is formed on a PCB and the feeding port cannot be put at the symmetrical center of the CSA. For these reasons, the prototype CSA has to be tuned in the exact frequency by adjusting its measurements. As a result, you get the final structure of a CSA as shown in **Figure 12** and **Table 2**. Note that PCB is used whose ε_r is 4.4, tanδ is 0.016, and thickness is 1.6 mm. S_{11} characteristics and radiation patterns of this CSA prototype are shown in **Figures 13(a)** and **(b)**, respectively. In **Figures 13(a)** and **(b)**, simulation results are compared with measurement results.

L_x	L_y	L_{y1}	L_{y2}	s_x (mm)	s_y (mm)	w_1 (mm)	w_2 (mm)	s_b (mm)	L_3	d (mm)
0.244 λ	0.628 λ	0.305 λ	0.281 λ	3.0	1.0	1.4	1.0	0.6	0.105λ	2.2

Table 2. The detailed measurements of the CSA prototype when the dielectric substrate is used whose ε_r is 4.4, tanδ is 0.016, and thickness is 1.6 mm.

These results show that CSA radiates CP waves at f_0 in which S_{11} reached −10 dB: the 3-dB axial ratio beamwidth is 100° and the 10-dB impedance bandwidth is 3%. There are some differences between the simulation and measurement results. These differences occur because the simulation results are calculated by feeding CSA through an ideal balanced port. On the other hand, measurement data are obtained by feeding CSA through a coaxial cable. These results show that the balance of the feeding port is important for CSA. Therefore, CSA is a simple CP antenna for application using balanced ports such as RFID tags.

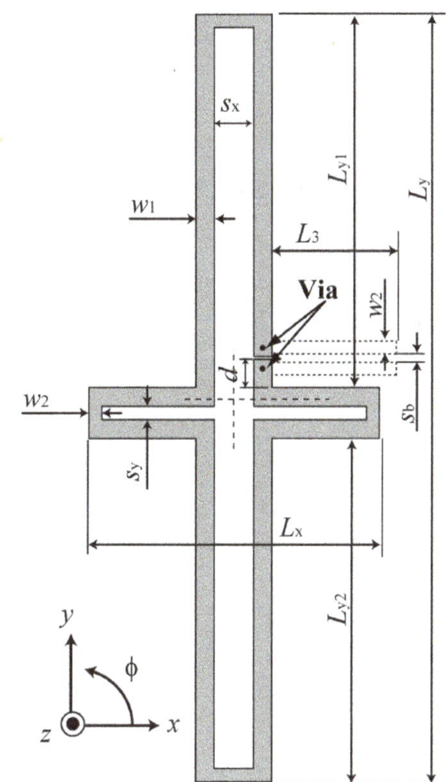

Figure 12. The structure of the CSA prototype.

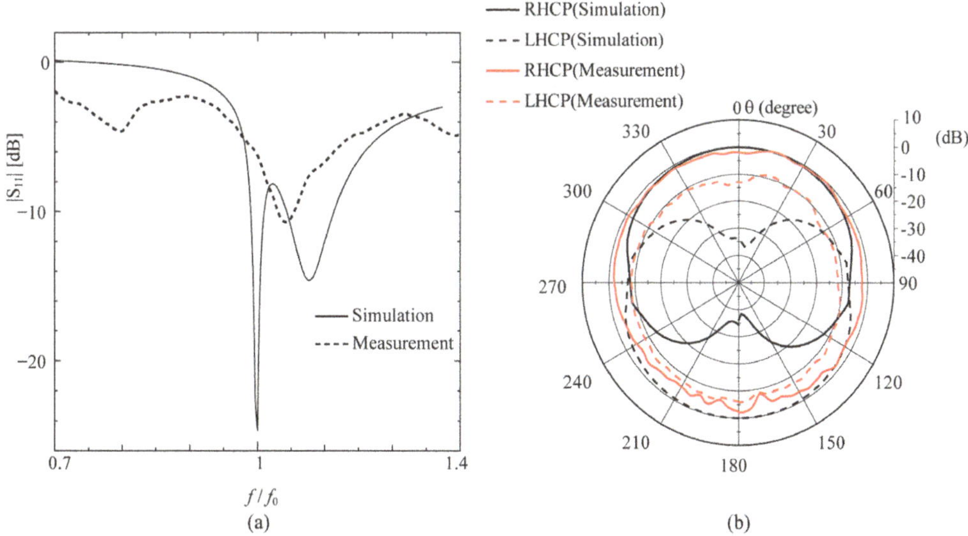

Figure 13. Characteristics of the CSA prototype as shown in **Figure. 12: (a)** S11 characteristics, and (**b**) radiation patterns at $f = f_0$ and $\phi = 0°$.

As a matter of fact, there are many applications in which coaxial cables are thought of as a useful feeding way. However, CSA needs to be fed through balanced ports. Measurement data in **Figure 13(b)** show that CSA no longer radiates CP waves, if it is fed through a coaxial cable without any treatments. In the next section, the feeding mechanism, which makes it possible to feed CSA through a coaxial cable, will be suggested. The feeding mechanism should be simple but helps CSA to radiate CP waves even if it is fed through a coaxial cable.

3. Dipole-fed cross-shaped spiral antenna

A simple way to feed CSA through a coaxial cable is discussed in this section. As mentioned before, the balance of the port is extremely important for CSA. For this reason, baluns, balanced-unbalanced transformers, are needed in general to feed CSA through a coaxial cable. However, baluns tend to need wider spaces than antennas. If the size of a balun becomes the same size as CSA, the usefulness of CSA's simple structure will be lost. Therefore, the feeding mechanism, which is as simple as possible, can be incorporated into CSA as its elements should be invented.

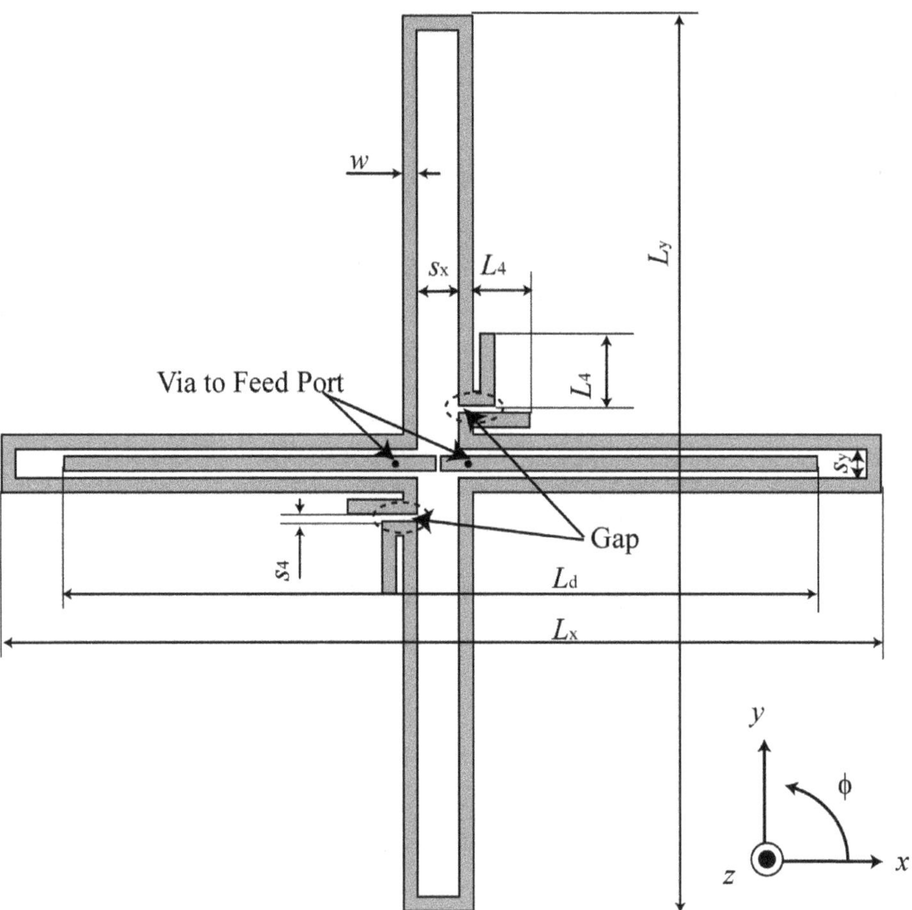

Figure 14. The structure of DFCSA.

Feeding loop antenna indirectly by using monopole elements suggested by Nakano is famous [9]. This is a feeding mechanism where a monopole element connected to a coaxial cable is put close to a loop antenna. Although this method could be used for CSA, it would need thick substrates and a ground plane. To keep CSA as a thin structure, the author tries to feed CSA by using a dipole antenna. The author also thought that a dipole antenna will work not only as a feeder but also as a radiator, if used for CSA.

In fact, putting a dipole antenna close to a loop antenna is difficult because they interfere with each other. This means that they no longer behave as they should. To solve this problem, gaps are made in CSA. **Figure 14** shows a representative structure of CSA fed with a dipole feeder. This antenna is named the dipole-fed cross-shaped spiral antenna (DFCSA). A dipole feeder is located in the center of CSA. The dipole feeder has a feeding port. CSA is turned around the dipole feeder in a cross shape. CSA does not have any feeding ports.

Next, two gaps are made at the CSA's joint between the upper element and the right element and between the lower element and the left element to avoid interference between the dipole antenna and CSA. S_{11} characteristics of DFCSA are shown in **Figure 15** when its measurements are set as shown in **Table 3**. S_{11} characteristics show that DFCSA has three resonant frequencies. Radiation patterns of each resonant frequency are shown in **Figure 16**. These results show that DFCSA radiates linear polarized (LP) waves at the lowest and highest frequencies and radiates CP waves at the middle frequency.

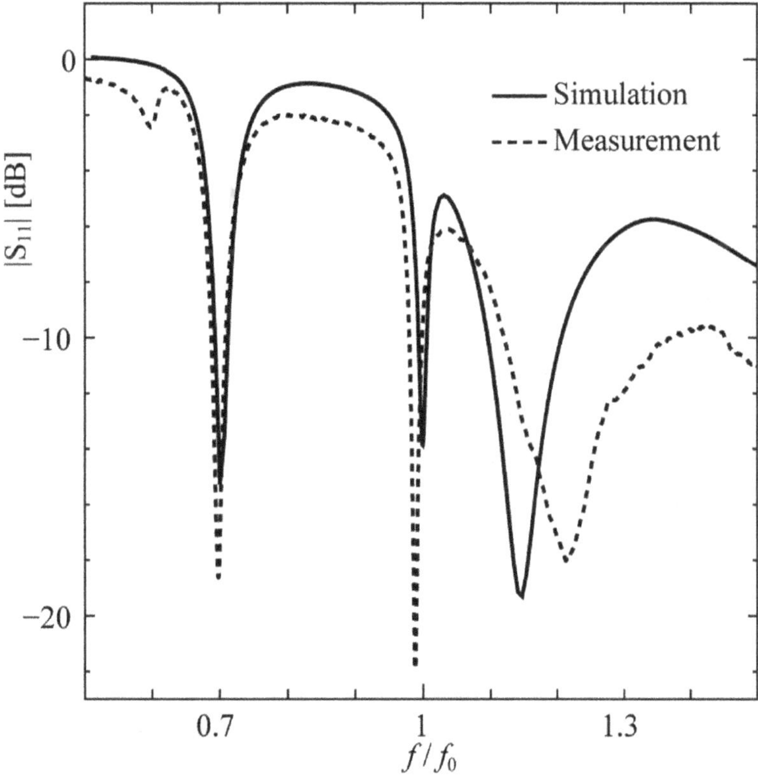

Figure 15. S_{11} characteristics of DFCSA.

L_x	L_y	L_d	s_x (mm)	s_y (mm)	w (mm)	s_4 (mm)	L_4 (mm)
0.668 λ	0.657λ	0.577λ	3.0	2.0	1.0	0.3	5.0

Table 3. The detailed measurements of the DFCSA when $\varepsilon_r = 4.4$, $\tan\delta = 0.016$, and thickness $t = 1.6$ mm of the dielectric substrate.

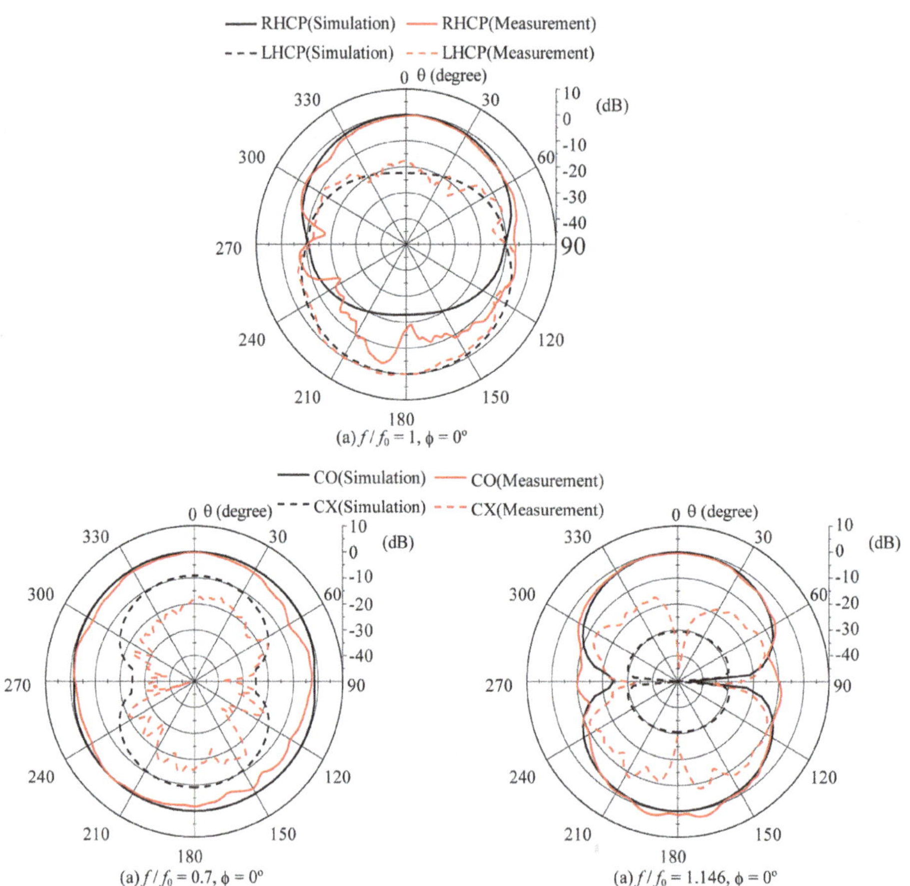

Figure 16. Radiation patterns at three resonant frequencies.

To validate the effectiveness of the dipole feeder, the measurement results are compared with the simulation results in **Figure 15** and **16**. In S_{11} results, there are some differences around the highest resonant frequency. This is because the dipole feeder works as a radiator around this frequency. Measurement radiation patterns show that DFCSA radiates CP waves at the middle resonant frequency and LP waves at the lowest and highest frequencies. For these reasons, the dipole feeder works well not only as a feeder for a coaxial cable but also as a radiator for LP waves.

To conclude, the mechanism of the dipole feeder is discussed. According to the structure of DFCSA, the dipole feeder seems to work like a microstrip balun constituted by the quarter-wavelength coupler of a dipole and CSA [10]. However, the reason why a dipole feeder helps feed CSA through a coaxial cable is simpler [11]. The measurement data in **Figure 16** show that radiation patterns of RHCP waves tilt by a few degrees. This is because CSA's right element is fed indirectly through the dipole's right element, which is connected to the outer conductor of the coaxial cable. In the same manner, CSA's left element is fed indirectly through the dipole's left element, which is connected to the center line of the coaxial cable. Although the current amplitude of the dipole's right element is different from that of the dipole's left element, there is no difference between the current phase of the right's dipole and that of the left's dipole. For this reason, the radiation directivity rotates some angles toward the direction of the right dipole element. This rotation angle is about 5°. The dipole feeder helps to reduce the effect of an unbalanced feed.

4. Concluding remarks

The simple CP antenna (CSA) and its principle were introduced. The fact that CSA was a printable antenna, which can be constituted on a single side of a circuit board without ground plane, was shown. When CSA was applied to antenna systems using balanced ports, it radiated good CP waves; the 3-dB axial ratio bandwidth was 1.5%, the 3-dB beamwidth was 100°, and the 10-dB impedance bandwidth was 3%. The maximum radiation gain and efficiency at the center frequency were, respectively, 0.96 dBi and 85% when the size of the antenna was about the square of the quarter wavelength.

An applied structure of CSA, which can be fed through a coaxial cable, was presented. The applied CSA was achieved by incorporating the dipole feeder into CSA. For this reason, the antenna was named the dipole-fed CSA (DFCSA). DFCSA radiated good CP waves without deterioration of S_{11} characteristics, even when it was fed through a coaxial cable. The merit of using the dipole feeder was that CSA developed into a multipolarization antenna. DFCSA had three resonant frequencies in S_{11} characteristics and radiated CP waves at one of them and LP waves at the others. The detailed characteristics of DFCSA are shown in **Table 4**.

Frequency f/f_0		1	0.7	1.146
Polarization		CP	LP	LP
Gain		−3 dBi	0 dBi	1.7 dBi
Efficiency		48%	76%	87%
Bandwidth	Axial ratio <3 dB	0.6%	–	–
	$S_{11} < -10$ dB	1.3%	2%	12%
Beamwidth		95°	Omnidirectional	100°

Table 4. Characteristics of DFCSA.

To my knowledge, CSA is the simplest CP antenna. It has multiple advantages: (1) it can be fed through not only balanced ports but also unbalanced ports, (2) it can radiate CP waves with a wide beamwidth, and (3) it can be developed into multiband and multipolarization antennas. It also has the large possibility of being flexibly modified to any structures and to be made electrically small in structure in future studies. The representative structures of CSA and DF-CSA are only some examples, which have not been miniaturized yet. They can be miniaturized by using shortening methods of dipole elements. Miniaturized structures, which can be called as electrically small ones, are going to be suggested in the author's next publications.

Acknowledgements

This work was supported by JSPS KAKENHI Grant Numbers 26420359 and 15H02135. The author would like to thank the Research Institute for Sustainable Humanosphere of Kyoto University.

Author details

Mayumi Matsunaga[1,2*]

Address all correspondence to: mmayumi@m.ieice.org

1 Department of Electrical and Electronic Engineering, Ehime University, Matsuyama, Ehime, Japan

2 Research Institute for Sustainable Humanosphere, Kyoto University, Uji, Kyoto, Japan

References

[1] Brown G. H.: The turnstile antenna. Electronics. 1936;9:14–17.

[2] Brown G. H.: Turnstile antenna. US Patent. 1941:2,267,550.

[3] Brown G. H. and Epstein J.: A pretuned turnstile antenna. Electronics. 1945;18:102–107.

[4] Lindenblad N. E.: Radio communication. US Patents. 1940:2,217,911.

[5] Lindenblad N. E.: Television transmitting antenna for empire state building. RCA Review 1939;3:387–408.

[6] Carver K. and Mink J.: Microstrip antenna technology. IEEE Transactions on Antennas and Propagation. 1981;29(1):2–24. DOI: 10.1109/TAP.1981.1142523.

[7] Matsunaga M.: A linearly and circularly polarized double-band cross spiral antenna. IEICE Transactions on Communications. 2016;E99-B(2):430–438. DOI: 10.1587/transcom.2015EBP3222.

[8] Bolster M. F.: A new type of circular polarizer using crossed dipoles. IRE Transactions on Microwave Theory and Techniques. 1961;9(5):385–388. DOI: 10.1109/TMTT.1961.1125358.

[9] Nakano H., Tagami H., Yoshizawa A. and Yamauchi J.: Shortening ratios of modified dipole antennas. IEEE Transactions on Antennas and Propagation. 1984;32(4):385–386. DOI: 10.1109/TAP.1984.1143321.

[10] Marchand N.: Transmission line conversion transformers. Electronics. 1944;17(12):142–145.

[11] Matsunaga M.: A dipole feeder for circularly and linearly polarized cross shape loop/spiral antennas. IEICE Electronics Express. 2016;13(12):20160426. DOI: 10.1587/elex.13.20160426.

3

Omnidirectional Circularly Polarized Antenna with High Gain in Wide Bandwidth

Bin Zhou, Junping Geng, Xianling Liang,
Ronghong Jin and Guanshen Chenhu

Abstract

A novel omnidirectional circularly polarized (CP) slot array antenna with high gain is proposed, which is based on the coaxial cylinder structure, and the orthogonal slots radiated the circular polarization wave around the cylinder. Further, the improved dual circularly polarized (CP) omnidirectional antenna based on slot array in coaxial cylinder structure is presented too, and two ports are assigned in its two side as left hand circularly polarized (LHCP) port and right hand circularly polarized (RHCP) port, respectively. The simulation and experiment results show their novelty and good performance of omnidirectional circular polarization with about 5 dBi gain in 5.2–5.9 GHz.

Keywords: omnidirectional circularly polarized (CP) antenna, slot array

1. Introduction

Omnidirectional circularly polarized (CP) antenna is a very good choice for wireless transmission and fast networking in wireless communication, which provides omnidirectional coverage, and is insensitive to the orientations of the waves. In addition, the circularly polarization will suppress the multipath interference in the complex environment [1].

In recent years, many kinds of omnidirectional CP antennas are investigated for the development of communication technology. In references [2–4], some planar microstrip omnidirectional CP antennas with dual-band are proposed, which are small, but the work bands are narrow. In references [5, 6], the omnidirectional CP dielectric resonator antennas (DRAs) are proposed, whose size is very small for the high permittivity, and their available impedance bandwidth can reach to 20% while the CP bandwidth is only 3–8% for the 3-dB axial-ratio

(AR). In reference [7], a CP antenna is designed which combines monopole and loop radiators to realize omnidirectional CP property, and the impedance bandwidth is about 15%. In reference [8], a compact size omnidirectional CP antenna is studied, in which four bended monopoles are simultaneously excited by feeding network. A broadband omnidirectional CP antenna is proposed in [9], of which the impedance bandwidth reaches to 45%, but its gain is not stable and high in the available bandwidth.

In another side, for a RF receiver which needs to receive electromagnetic signals with any polarizations mode and from any directions on the ground, the dual CP omnidirectional antenna is also very significant and valuable. The dual CP omnidirectional antenna is widely used, such as in wireless communications, radio broadcasting, and navigation radar [1, 11, 12]. In recent years, there are many researches about omnidirectional antenna and CP antenna. In reference [3], a miniaturization of omnidirectional CP antenna relies on folding the antennas patch underneath itself, decreasing overall footprint, which is small, but with a narrow bandwidth of only 0.6% (2.392–2.407 GHz) and the axial ratio in some directions of the omnidirectional plane is higher than 3 dB. For the antenna proposed in reference [13], by introducing several inclined slits to the diagonal and sidewalls of the rectangular dielectric resonator (RDR) and also deducting a rectangular part of the top wall of the linearly polarized (LP) rectangular dielectric resonator antenna (RDRA), degeneracy modes are excited to generate the circularly polarized (CP) fields, while the bandwidth is narrow and the gain fluctuation in the omnidirectional plane is higher than 5 dB from the results of the radiation patterns, so it does not perform well in the omnidirectional character. Literature [14] reports a dual CP antenna which is excited at the second-order mode to generate the conical radiation pattern and is fed by a hybrid coupler to obtain the dual CP operation, but it realizes a conical beam instead of an omnidirectional coverage radiation pattern. There are only a few researches about omnidirectional dual CP antenna. In literature [15], an omnidirectional dual-band dual circularly polarized microstrip antenna is proposed, while the dual circularly polarization means that the antenna provides different single RHCP and LHCP in its two different bandwidths, respectively, so it is single circularly polarized antenna in the specific frequency bandwidth actually. In reference [16], a dual-CP omnidirectional antenna is proposed, which consists of four tilted dipoles with parasitic elements for each sense of circular polarization.

To achieve compact omnidirectional CP antenna and to overcome the shortcomings of previous antenna listed above, here we design two kinds of omnidirectional circularly polarized slot array antenna, one is single CP polarized antenna, and the other one is the dual CP omnidirectional antenna. So in this chapter, the antenna design includes two parts:

1. An omnidirectional circularly polarized slot array antenna with high gain in a wide bandwidth.

2. Dual circularly polarized omnidirectional antenna with slot array on coaxial cylinder.

In the first part of this chapter, we propose our idea of omnidirectional circularly polarized slot array antenna, continually, introduce the design principle of this CP antenna, and the method to design the omnidirectional CP antenna. Later, an example is given, and the experimental results show its novelty and good performance.

In the second part of this chapter, we introduce the antenna with two ports in its two sides to realize dual circular polarization, and these two ports are assigned in its two side as left hand circularly polarized (LHCP) port and right hand circularly polarized (RHCP) port, respectively. The design principle is much the same with the first antenna presented but much more simple and symmetric. And detailed mathematical derivation is presented to prove the antenna's circular polarization property. Same as the first part of this chapter, results and discussion are presented in the final of the section to show its novel performance.

2. An omnidirectional circularly polarized slot array antenna with high gain in a wide bandwidth

In this section, a novel omnidirectional CP antenna based on the coaxial cylinder waveguide with slots array around the cylindrical conductor shell is designed. The CP wave is radiated by the four orthogonal slots pairs with a wavelength interval around the cylinder shell, and combined the omnidirectional CP wave. Four ring slot pairs along the axis construct a slot array around the cylinder shell that achieves the stable high gain. Furthermore, it is convenient to increase or reduce the number of the basic slot array elements along the axis to adjust the antenna gain for wider application.

2.1. Antenna structure

The structure of the proposed omnidirectional CP antenna is shown in **Figure 1**. It is based on the slots array in the coaxial cylinder, and the interval of the inner and outer conductors is filled by Teflon. The thickness of the outer cylinder conductor is t = 1.5 mm. The antenna includes three parts: radiation slot array, impedance matching part, and feeding part.

The orthogonal slot pairs are the basic radiation elements that are symmetrically distributed around the cylinder conductor shell, as shown in **Figure 1**. These elements will radiate the omnidirectional CP wave. Four of the basic radiation elements group along the axis constructs the slot array.

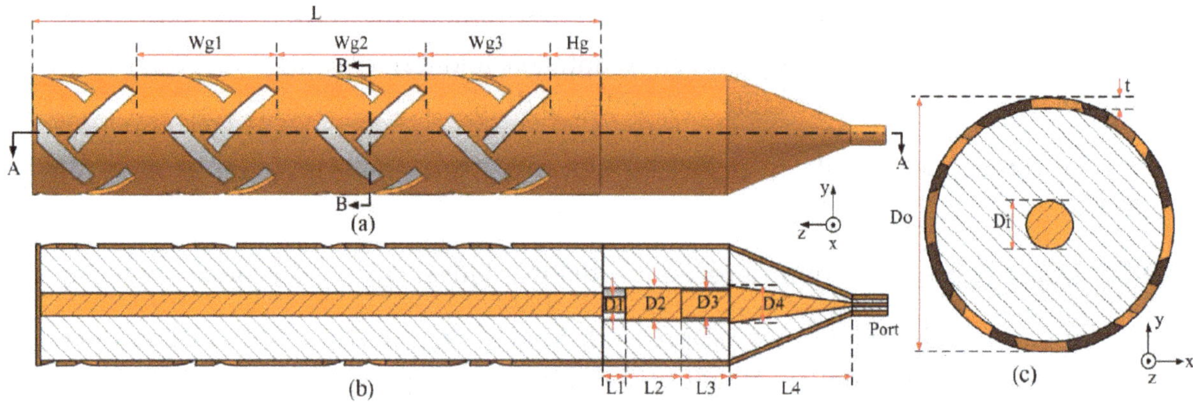

Figure 1. Geometry of the omnidirectional CP antenna. (a) Side view. (b) A-A section view. (c) B-B section view.

In **Figure 1(b)**, three coaxial cylinders L1, L2, and L3 form the impedance matching part. In order to install the small SMA adaptors with a characteristic impedance of 50 Ω to the large-size radiation part with an input impedance varying with the frequency, the coaxial cylinders L1, L2, and L3 with optimized characteristic impedance and electrical length are designed based on the theory of transmission line.

The feeding part L4 is a coaxial tapered line with stable characteristic impedance by keeping a constant ratio of outer and inner radius, which will connect the small-size SMA adaptor to the large-size matching part.

2.2. Design principle

The geometry structure of the slot pairs are shown in **Figure 2**. One slot tilts at 45°, while the other tilts at -45°. Moreover, they are placed with interval $\lambda_g/4$ (λ_g is the medium wavelength) distance along the axis, and the waves radiated from the slot pairs are orthogonal to each other with 90° phase difference, so that they achieve the RHCP property [10]. In another side, the slot is equivalent to a magnetic current source like a dipole tilted at an angle of 45°, the slot is designed with length $L_s = 2\lambda_g/2$.

Four of the slot pairs uniformly distributed around the cylinder conductor shell can be regarded as the basic omnidirectional CP element which is shown in **Figure 1(c)**. For high-gain, four rounds of the basic elements along the axis construct the slot array in **Figure 1(b)**.

The four rounds of elements are placed at an interval distance $W_g = \lambda_g$, so that the waves radiated from every element of the slot array keep the same phases. The filled substrate is Teflon with permittivity $\varepsilon_r = 2.1$, so the interval between the slot array elements is $\lambda_g = 0.7$

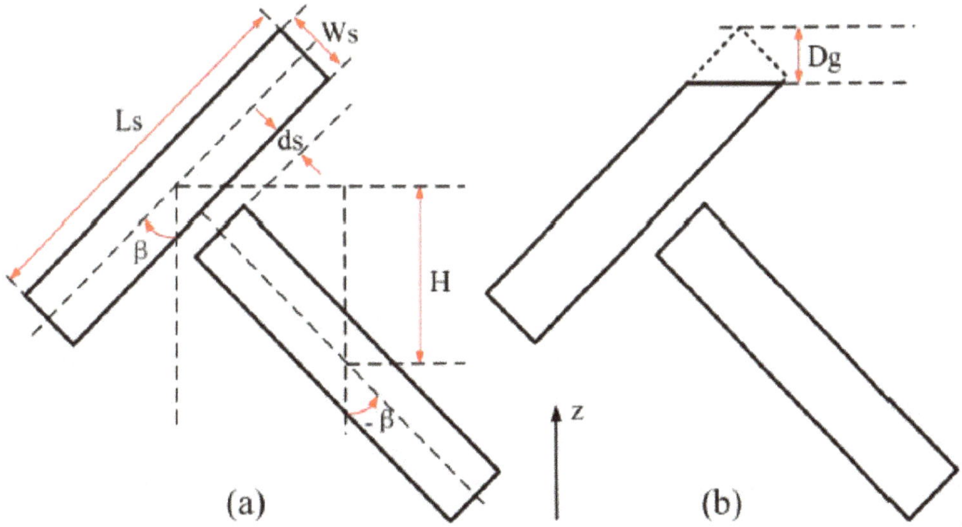

Figure 2. Geometry of slot pairs on outer conductor of the coaxial cylinder. (a) Normal slot pairs. (b) Changed slot pairs.

λ_0 (λ_0 is the space wavelength). While the slot element close to the end is unique as shown in **Figure 2(b)**, it is cut at a length as shown in **Figure 2(a)**. The end element is practically like a matched load to radiate electromagnetic wave and relieves the reflected wave in the shortened terminal of the coaxial cylinder, which ensures the traveling wave running in this structure and improves radiation efficiency of the antenna. So, the antenna performs stable high gain in a wide CP and impedance bandwidth.

2.3. Simulate verification

The omnidirectional CP antenna is simulated and optimized using CST Microwave Studio software based on the finite integration technique (FIT). The optimized geometric parameters are summarized in **Table 1**.

D_o	32 mm	D4	10 mm
D_i	6 mm	L1	6.1 mm
L	161.1 mm	L2	15.8 mm
W_g1	39.59 mm	L3	13.2 mm
W_g2	42.27 mm	L4	40 mm
W_g3	35.77 mm	Ls	23.7 mm
Hg	14 mm	Ws	4.5 mm
D1	4.2 mm	H	11.8 mm
D2	8.7 mm	Dg	3.6 mm
D3	7.6 mm	ds	1.65 mm
β	45°		

Table 1. Optimized geometric parameters of the omnidirectional CP antenna.

Figure 3 shows the magnetic field distribution of the antenna at t = 0, T/4, T/2, 3T/4. In order to clearly explain the working principle of the antenna, the magnetic fields on the A-A section and B-B section of the antenna are shown in **Figure 3**. The magnetic fields on the B-B section of the antenna in turn at t = 0, T/4, T/2, 3T/4 display how the omnidirectional CP waves are formed by the basic omnidirectional CP elements in a whole period time (T). The magnetic fields on the A-A section of the antenna detailed shows that the radiated waves always keep the same phase in every slot element. Therefore, the cylindrical slot array antenna performs a stable high gain in the available bandwidth.

At $t = 0$, the magnetic fields are strongest and outward in each slot close to the feeding port, while the fields in the end slot is very weak, as shown in **Figure 3(a)**. At this time, every slot radiates a 45° linear polarized wave. At $t = T/4$, the magnetic fields are the strongest and in outward direction in every slot close to the terminal, while the fields in the slot near the feeding port is very weak as shown in **Figure 3(b)**. At this time, every slot radiates a -45° linear polarized wave. At $t = T/2$ and $t = 3T/4$, the magnetic fields vary in the opposite direction as shown in **Figure 3(c)** and **(d)**, respectively. Thus, there is always a 90° phase difference between the 45° linear polarized wave and the -45° linear polarized wave in all of the slot pairs around the coaxial cylinder.

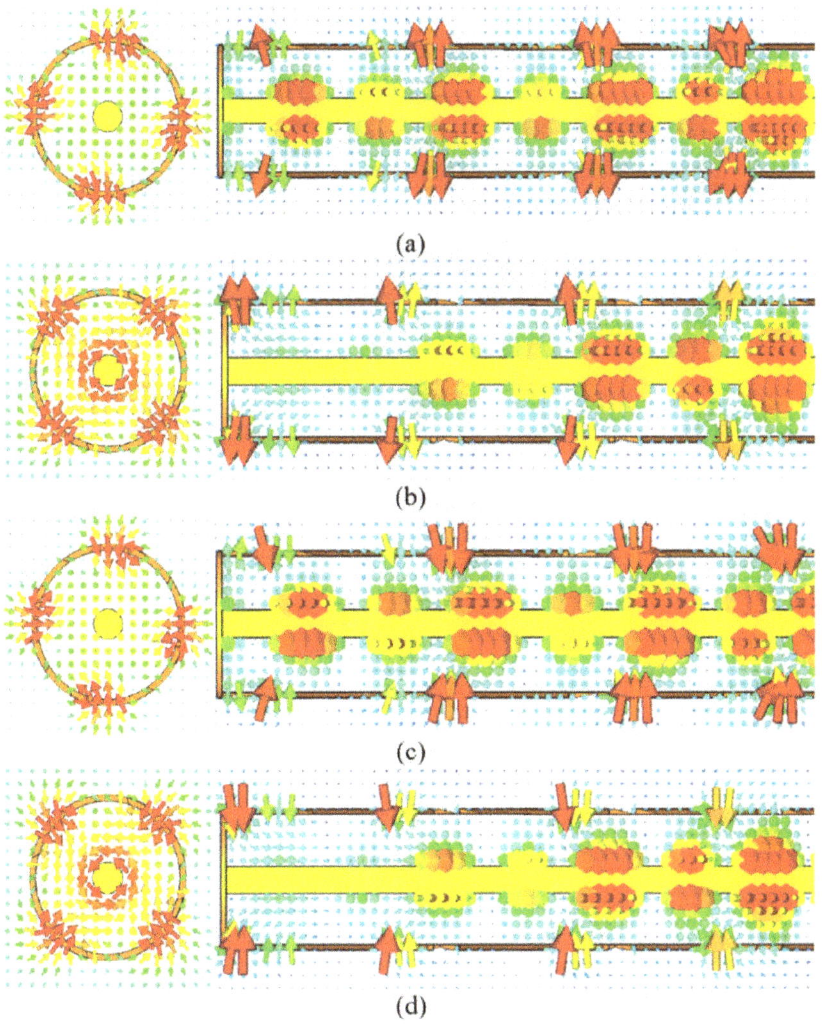

Figure 3. Magnetic fields distribution in (right) A-A section and (left) B-B section of the antenna. (a) $t = 0$. (b) $t = T/4$. (c) $t = T/2$. (d) $t = 3T/4$.

2.4. Results and discussion

Simulated and measured results of the proposed antenna are presented in this section. **Figure 4** shows the test scenario of the omnidirectional CP antenna.

Figure 4. Photograph of the omnidirectional CP antenna being tested in the anechoic chamber.

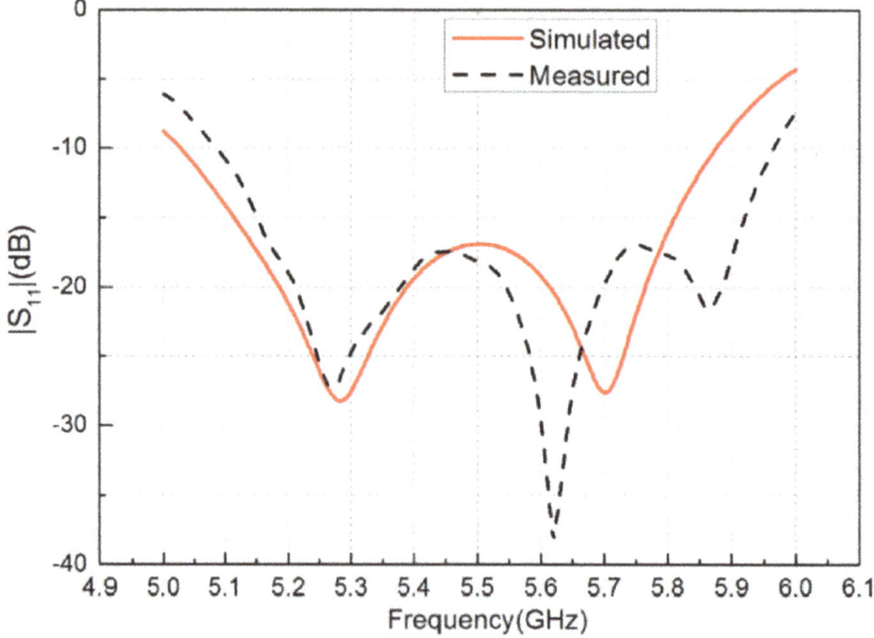

Figure 5. Measured and simulated of the omnidirectional CP antenna.

The simulated and experiment measured return loss are shown in **Figure 5**. It can be observed that the measured impedance bandwidth of $S_{11} < -10$ dB is 16.4% ranging from 5.05 to 5.95 GHz. The measured result shows the same trend with the simulated result, but slightly deflected to the high frequency. It may be because the medium of Teflon is impure, and the real permittivity value of ε_r is smaller than the simulated value of $\varepsilon_r = 2.1$. The simulation and experiment average axial ratio and gain in the horizontal plane (*xy*-plane) is drawn in **Figure 6**. The experiment measured gain varies from 5 to 7 dBic in 5.1–5.9 GHz, and the average ARs below 3 dB is in 5.1–5.9 GHz. Note that the gains of the experiment are 0.8 dB smaller than simulated near the center frequency, and the average ARs are some worse than the simulated results. Besides the imprecise dialectical constant and the experiment error, the reason for these differences between measured and simulated results is mainly attributed to the manufacturing error because the performance of the antenna is sensitive to the slots size. Generally, the proposed antenna achieves the stable high gain within the available bandwidth. **Figure 7(a)–(c)** shows the simulated RHCP and left-hand CP (LHCP) normalized radiation patterns in the omnidirectional plane (*xy*-plane), E-plane (*yz*-plane), and normalized experimental radiation patterns at 5.2, 5.5, and 5.8 GHz, respectively. In **Figure 7**, the cross-polarization in the omnidirectional plane is lower by 16 dB than the copolarization, and the radiation efficiency is 94% in the operating band 5.1–5.9 GHz. The measured and simulated results are very consistent, and the variation in the omnidirectional plane is very small, which means that the proposed antenna has good omnidirectional characteristics.

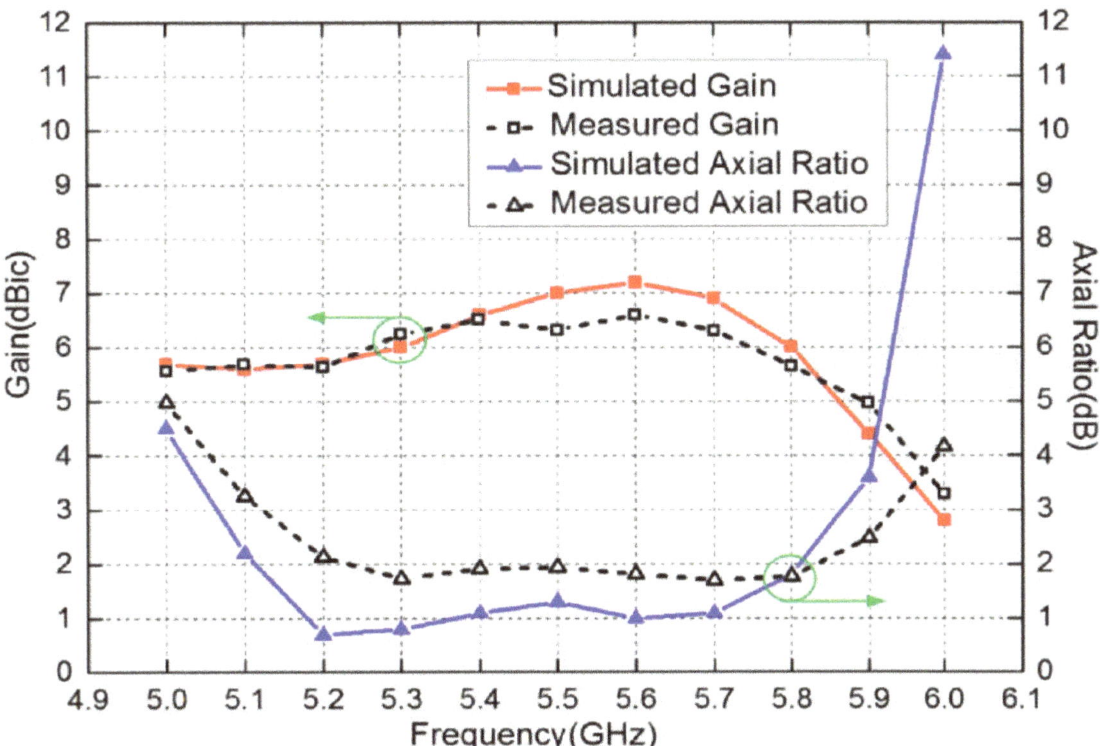

Figure 6. Measured and simulated gain and average axial-ratio results of the omnidirectional CP antenna.

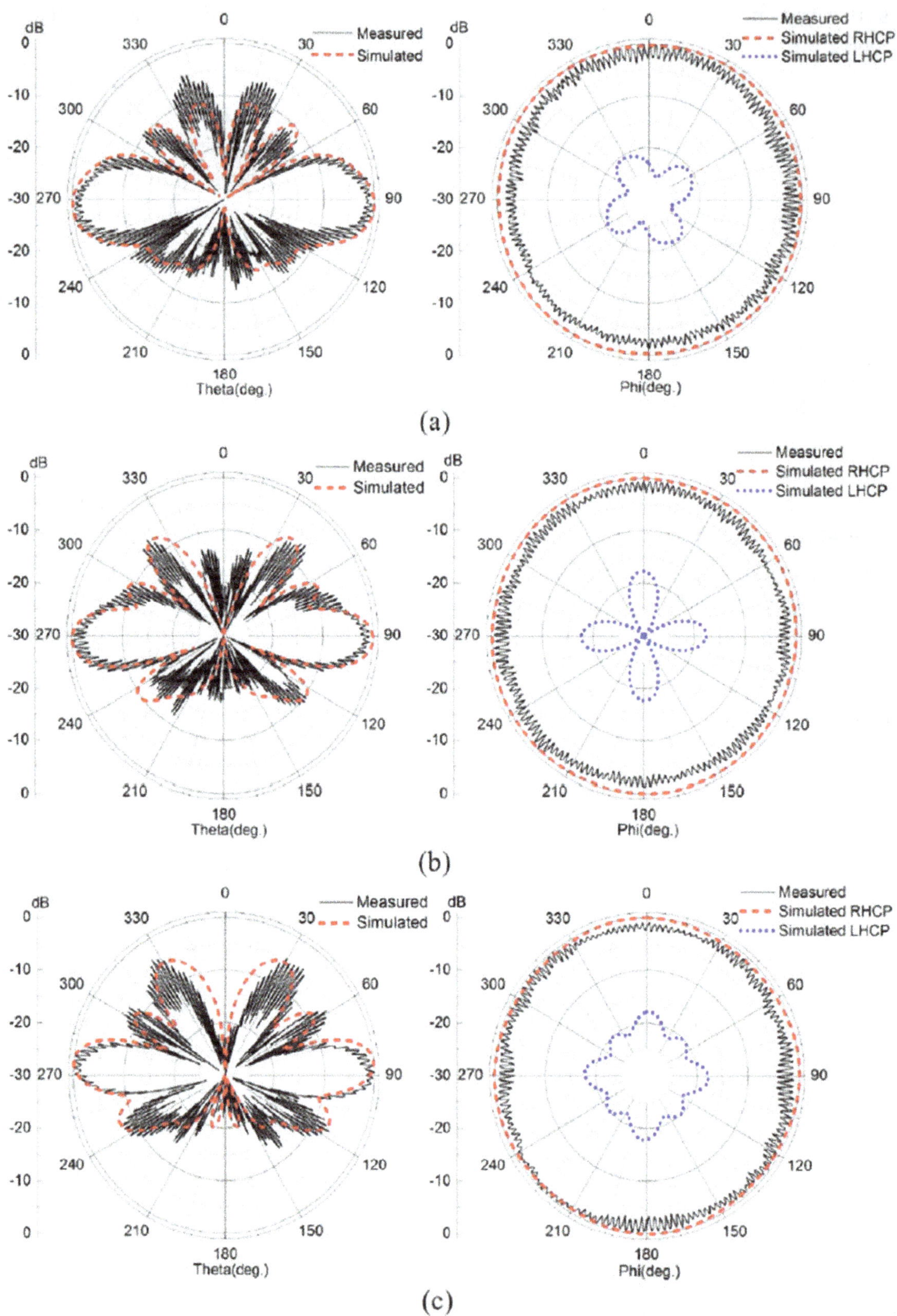

Figure 7. Simulated RHCP and LHCP normalized radiation patterns and measured normalized radiation pattern in the E-plane (yz-plane on the left) and the omnidirectional plane (xy-plane on the right). (a) 5.2 GHz. (b) 5.5 GHz. (c) 5.8 GHz.

3. Dual circularly polarized omnidirectional antenna with slot array on coaxial cylinder

Further, we introduce the dual circularly polarized omnidirectional antenna which has two ports in its two sides to realize dual circularly polarized property. The two ports are assigned in its two sides as left hand circularly polarized (LHCP) port and right hand circularly polarized (RHCP) port, respectively. The proposed antenna achieves a bandwidth of 16.4% ranging from 5.05 to 5.95 GHz with an isolation higher than 15 dB between the two CP ports, and the return loss (RL) is lower than -10 dB within the bandwidth in both of the two ports. From the measured results, the average axial ratio (AR) of the proposed antenna in omnidirectional plane is lower than 1.5 dB.

3.1. Antenna structure

As shown in **Figure 8**, the proposed dual omnidirectional CP antenna is based on the single CP omnidirectional antenna represented in literature [17] constructed with coaxial cylinder structure. Two ports are assigned in the two sides of the coaxial cylinder as left hand circularly polarized (LHCP) port and right hand circularly polarized (RHCP) port, respectively. The characteristic impedance of the coaxial cylinder is designed as 50 Ω in this section, which is the same as the characteristic impedance of the two SMA adaptor ports. The coaxial tapered line is in both of the two sides between the slot arrays and the two SMA adaptor ports, and it achieves the goal of port match.

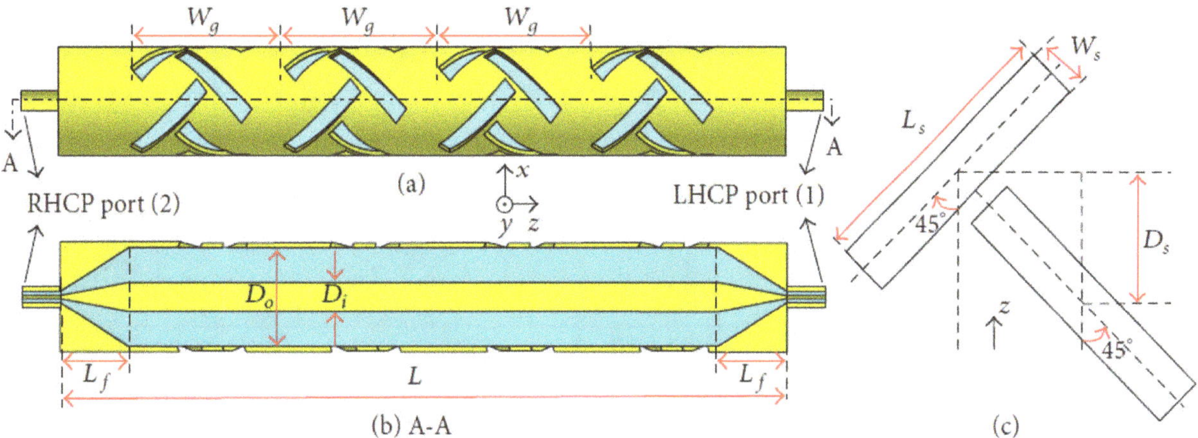

Figure 8. Geometry of the omnidirectional dual circularly polarized (CP) antenna. (a) Side view. (b) A-A section view. (c) Geometry of the slot pairs.

The basic radiation elements are slots in the out conductor of the coaxial cylinder. There are four circles of slots along the axis of the coaxial cylinder. Each circle of the slots includes 8 slots which are perpendicular to each other and symmetrically arranged around the coaxial cylinder axis in the out conductor. Every circle of the slots can be seen as four perpendicular slot pairs around the axis.

The structure of the perpendicular slot pairs is shown in **Figure 9**. Each of the slots leans at an angle of 45° with the axis of the coaxial cylinder and the interval of the two perpendicular slots is $\lambda g/4$ (λg is the medium wavelength) along the feed direction. In that case, the electric fields radiated from the slot pairs are vertical with each other and have a 90° phase difference, so the circularly polarized wave is generated. For the distance of the $\lambda g/4$ between the adjacent pairs of slots, the reflection power from radiation slots back to the feeding point is cancelled. As a slot antenna, the slot in the out conductor of the coaxial cylinder can be equivalent as magnetic dipole with a length of $\lambda g/2$ in the vertical direction with the axis. Four of the slot pairs symmetrically surrounding the coaxial cylinder axis achieve the omnidirectional coverage character. Four circles of the slot pairs, which are identical with each other, are arranged along the axis to constitute the slot array which will provide a higher gain. To achieve the omnidirectional pattern, the electromagnetic wave radiated from every circle of slots should be of the same phases, so the four circles of the slot are arranged with a distance of λg.

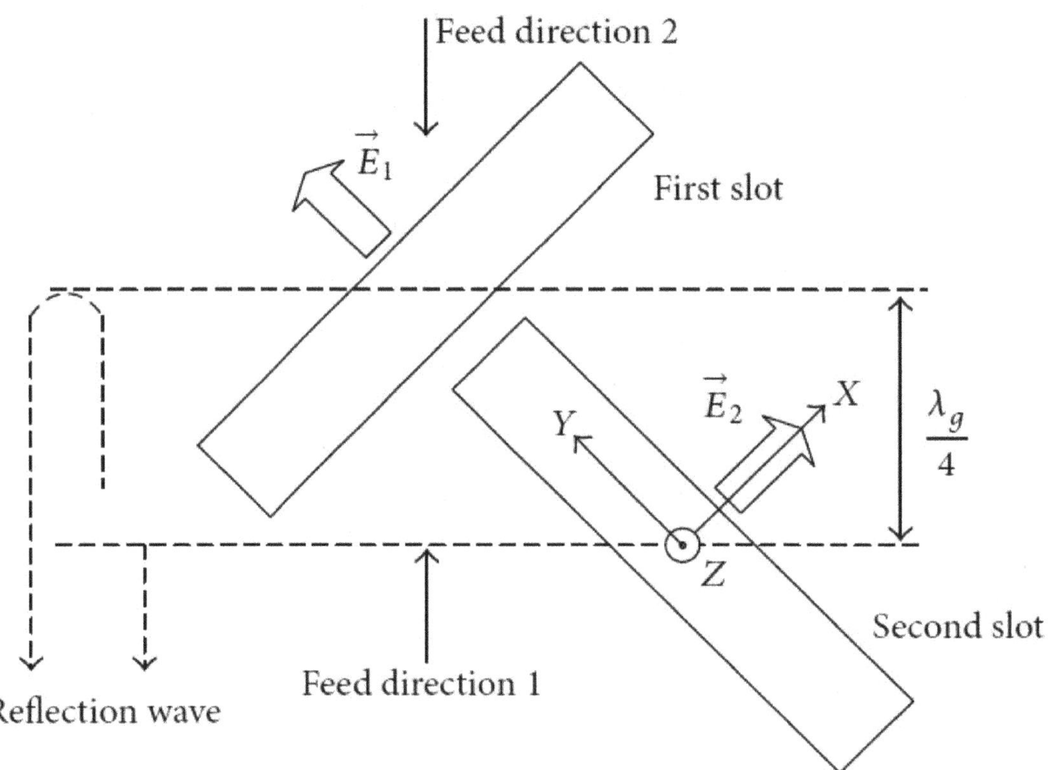

Figure 9. The adjacent perpendicular slot pairs.

It is a symmetrical structure antenna. The thickness of the out conductor of the coaxial cylinder is t = 1.3 mm, and the space between the out conductor and the inner conductor of the coaxial cylinder is filled with Teflon medium with a permittivity of ε_r = 2.1. The proposed antenna is simulated and optimized by CST and the optimized geometric parameters are summarized in **Table 2**.

L	153.5 mm	Wg	40 mm
D_i	7.6 mm	Do	28 mm
L_s	24.4 mm	Ws	4.3 mm
D_s	11.8 mm	Lf	18 mm

Table 2. Optimized geometric parameters of the omnidirectional CP antenna.

3.2. Design principle

Figure 9 shows two adjacent slots from all the slots in the out conductor, which are perpendicular to each other. The reference coordinate system is created with the slot pairs, whose y- and x-axes are of the same direction with the polarization of the electric fields radiated by the two slots, respectively, and the $+z$-axis is the same direction with the propagation direction of the electric fields [10, 18]. It is assumed that the total electric fields are $\vec{E}$ and the fields radiated from the first slot and the second slot are $\vec{E}_1$ and $\vec{E}_2$, respectively. Then their relation is

$$\vec{E} = \vec{E}_1 + \vec{E}_2 = E_x\vec{a}_x + E_y\vec{a}_y \tag{1}$$

For the distance of the two slots, it is supposed that the amplitude of the E_x and E_y is E_{x0} and E_{y0} ($E_{x0} > 0$, $E_{y0} > 0$), and compared to E_x, E_y is laggard in phase with a difference of ϕ. Then Eq. (1) can be changed as

$$\vec{E}(z,t) = E_{x0}\cos(wt - kz)\vec{a}_x + E_{y0}\cos(wt - kz - \phi)\vec{a}_y \tag{2}$$

where

$$\begin{aligned} E_x(z,t) &= E_{x0}\cos(wt - kz)\vec{a}_x \\ E_y(z,t) &= E_{y0}\cos(wt - kz - \phi)\vec{a}_y \end{aligned} \tag{3}$$

The relation between $E_x(z, t)$ and $E_y(z, t)$ can be concluded from Eq. (3):

$$\left[\frac{E_x(z,t)}{E_{x0}}\right]^2 + \left[\frac{E_y(z,t)}{E_{y0}}\right]^2 - \frac{2E_x(z,t)E_y(z,t)}{E_{x0}E_{y0}}cos\phi = \sin^2\phi \tag{4}$$

Since the distance of the two slots in the feed direction is $\lambda_g/4$, the phase difference is $\phi = \pm 90°$. It is supposed that the amplitude of electric fields is $E_{x0} = E_{y0} = E_0$, then from Eq. (4):

$$E_x^2(z,t) + E_y^2(z,t) = E_0^2 \tag{5}$$

It can be seen from Eq. (3) that the amplitude of $\vec{E}(z, t)$ will not change with time t, and the angles between the polarization direction of $\vec{E}(z, t)$ and $+x$-axis direction are

$$\alpha = \tan^{-1}\frac{E_0\cos\left(wt - kz \pm \frac{\pi}{2}\right)}{E_0\cos(wt - kz)} = \tan^{-1}[\pm\tan(wt - kz)] = \mp(wt - kz) \tag{6}$$

In any positions, z is a constant, and the polarized vector direction of the field $\vec{E}(z, t)$ is rotary with a constant angular frequency ω by the increasing of t. As shown in **Figure 11**, when excited as fed direction 1, the phase of E_y is laggard compared to the phase of E_x, so $\phi = 90°$ and $\alpha = \omega t$. The electric fields vector direction is anticlockwise, so it is right hand circularly polarized (RHCP) wave. When excited as fed direction 2, the phase of E_y exceeds the phase of E_x, so $\phi = -90°$ and $\alpha = -\omega t$. The electric fields vector direction is clockwise, so it is left hand circularly polarized (LHCP) wave. Comparing **Figure 11** with the slots distribution in **Figure 11**, the RHCP port and LHCP port are corresponding with fed direction 1 and fed direction 2, respectively.

For the different distances of the two slots relative to the fed port, the electric fields amplitude of the two slots is unequal $\left(E_{x0} \neq E_{y0}\right)$. This is an influence factor of the axial ratio and their relation is reflected by the axial ratio character of the antenna.

3.3. Results and discussion

Simulated and measured results of the proposed antenna are presented in this section. To measure the LHCP property and RHCP property of the antenna, there are two steps. When port 1 (LHCP port) is excited and port 2 (RHCP port) is terminated with a 50 Ω load, the antenna generates the LHCP radiation and we get LHCP results, whereas when port 2 is excited and port 1 is terminated with a 50 Ω load, the antenna generates the RHCP radiation and we get RHCP measurement results. **Figure 10** shows the antenna structure and the test scenario of the omnidirectional dual CP antenna.

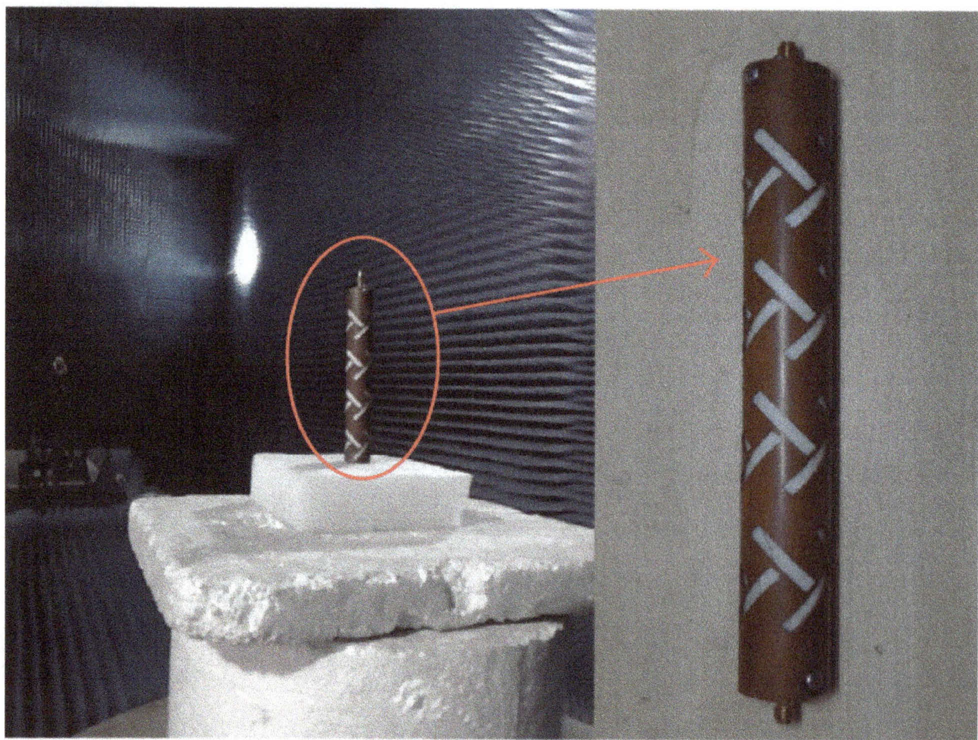

Figure 10. Photograph of the omnidirectional dual CP antenna being tested in the anechoic chamber.

The measured *S*-parameters results are shown in **Figure 11**. The proposed antenna achieves a bandwidth of 16.4% at 5.05–5.95 GHz, during which the bandwidth isolation between the two ports is higher than 15 dB, and both of the measured |*S*11| and |*S*22| values are lower than -10 dB. The reflection of the two ports is small and the isolation between the two ports is high, which reflects that the antenna works well with two polarizations.

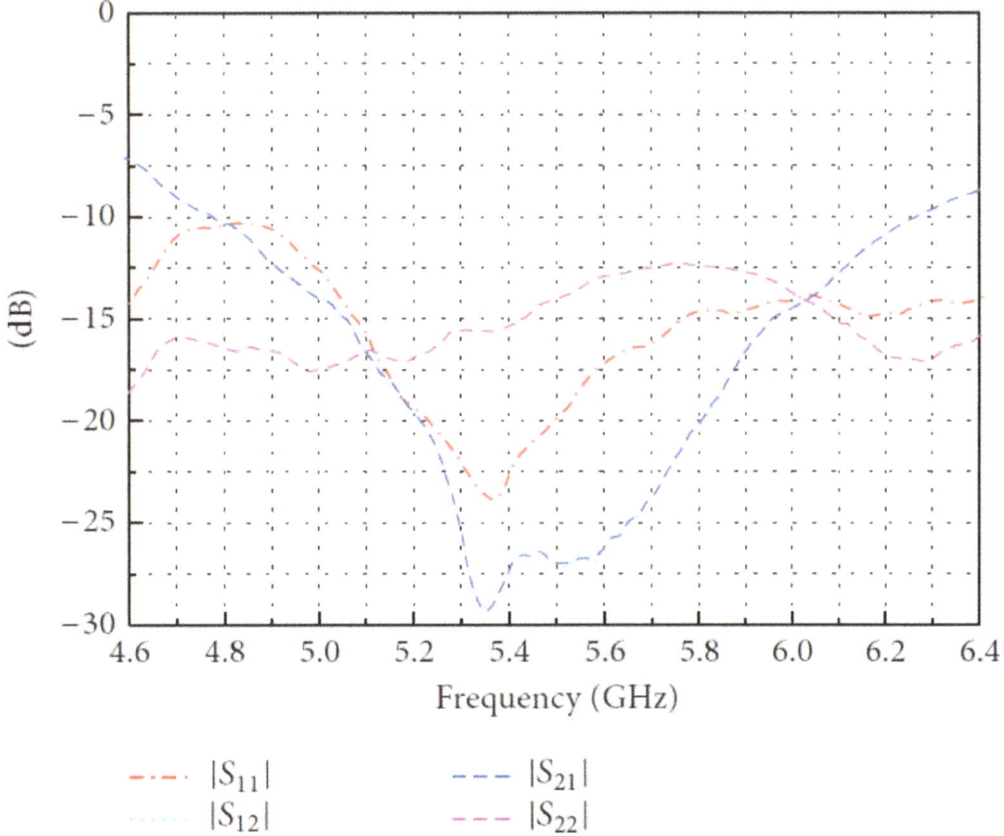

Figure 11. Measured *S*-parameters results.

Figure 12 shows the simulated and measured normalized copolarization and cross-polarization radiation patterns when excited RHCP port and LHCP port, respectively, are in the *xy*-plane (omnidirectional plane) and *xz*-plane at 5.1, 5.5, and 5.9 GHz. From the measured radiation pattern results of the *xy*-plane, the proposed antenna performs a good omnidirectional character.

It is noted that the measured radiation patterns have the same trend with the simulated radiation patterns, but there are still a few discrepancies between measured results and simulated results. The reasons for this difference are mainly because of the machine error and the imprecision of the medium. It can be also seen from cross-polarization results that the cross-polarization is more than 15 dB lower than the copolarization in both *xy*-plane and *xz*-plane. The measured axial ratio patterns that only excited LHCP port in *xy*-plane and *xz*-plane at 5.1, 5.5, and 5.9 GHz are shown in **Figure 13**. It can be seen that the axial ratio is below 3 dB in the omnidirectional plane within the operation bandwidth. Comparing the radiation patterns and

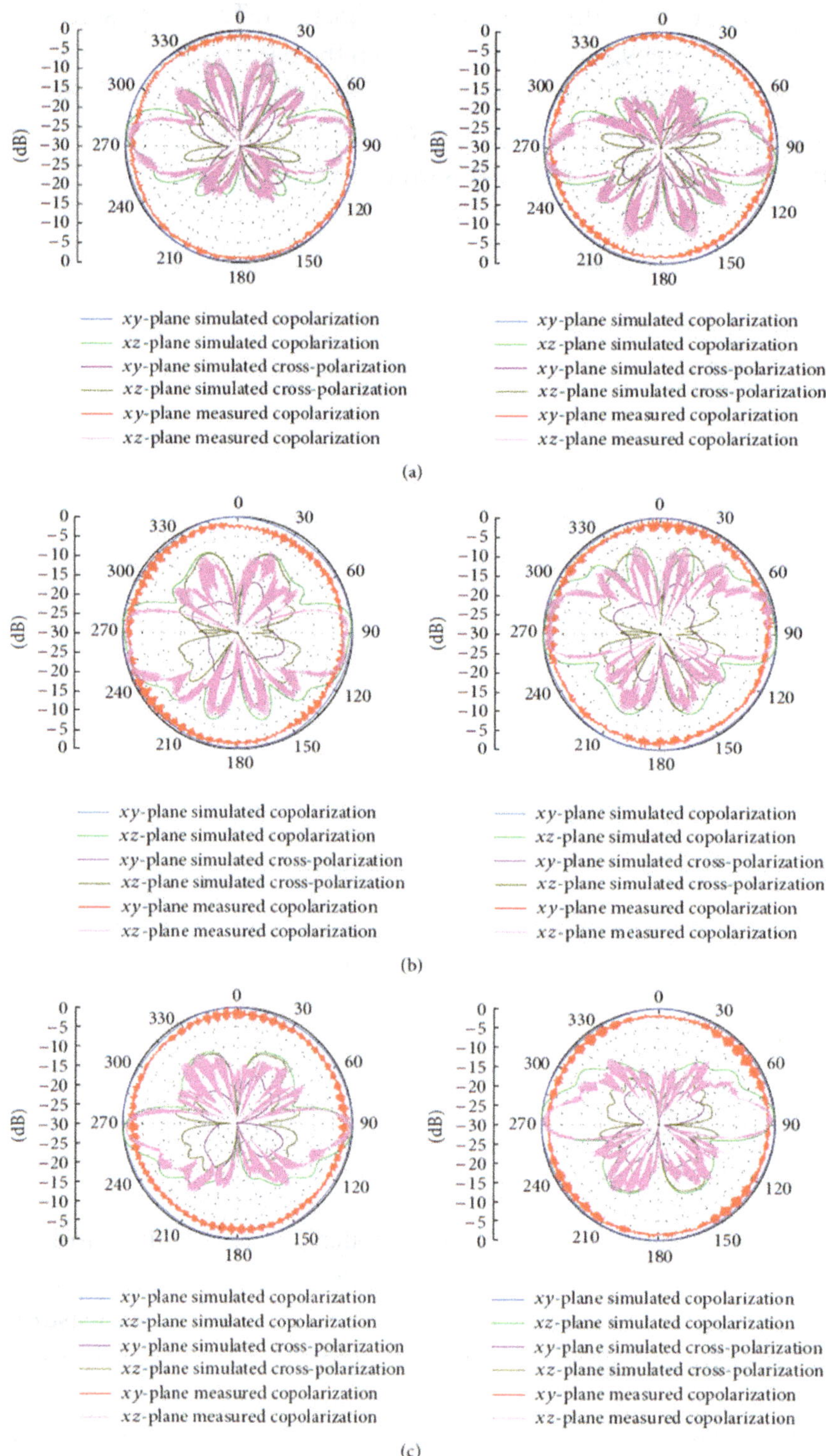

Figure 12. Simulated and measured LHCP (in left) and RHCP (in right) normalized copolarization and cross-polarization radiation patterns in the *xy*-plane (omnidirectional plane) and *xz*-plane. (a) 5.1 GHz, (b) 5.5 GHz, and (c) 5.9 GHz.

axial ratio patterns results, the axial ratio is lower than 3 dB during the half power beam width. When only excited the RHCP port, the axial ratio is below 3 dB in the omnidirectional plane which is similar as show **Figure 13**. It can be noted that the axial ratio patterns in the *xy*-plane are small ripples. From the generation principle of the dual CP wave as analyzed in Section

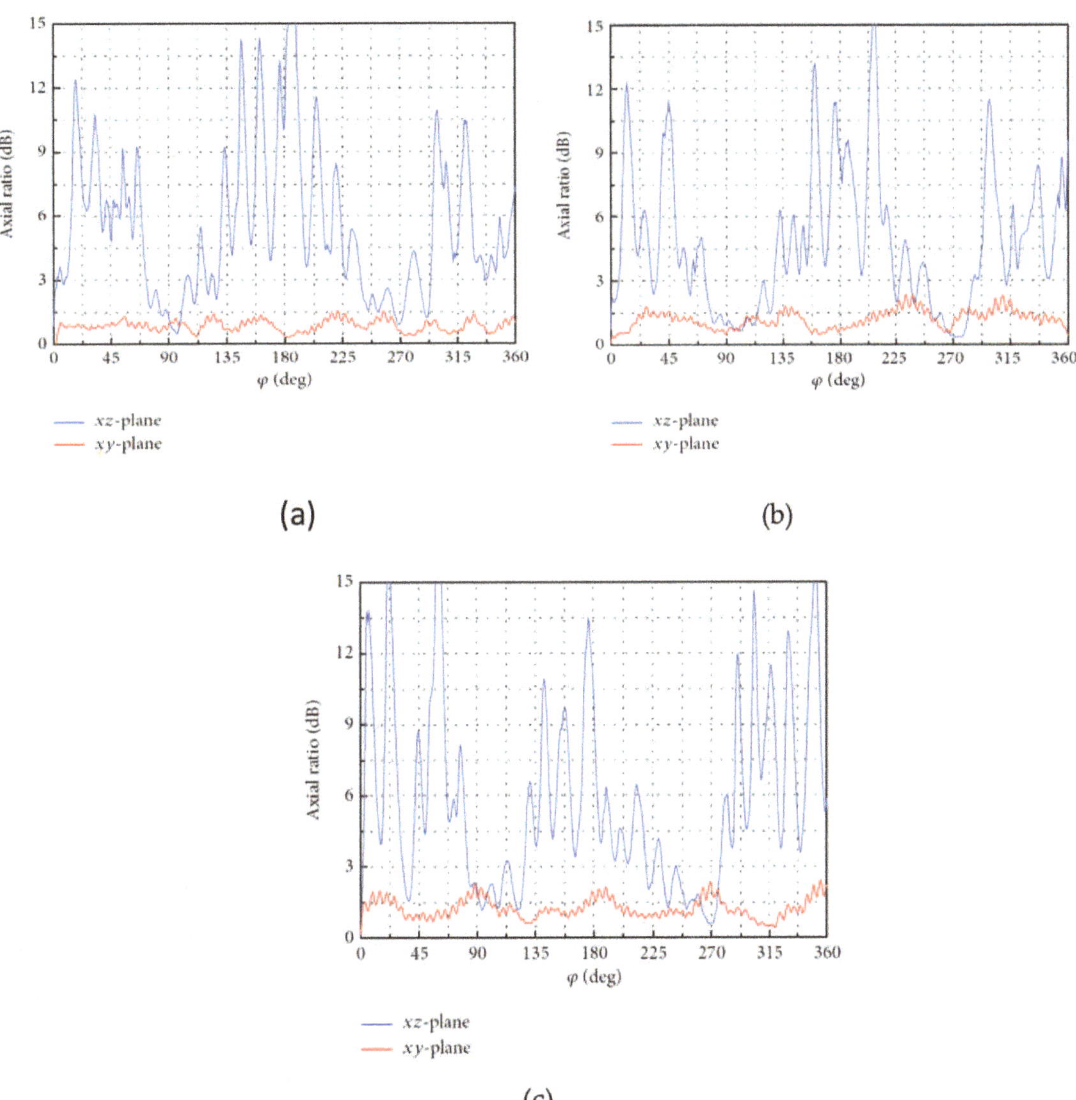

Figure 13. Measured axial ratio patterns in xy-plane and xz-plane. (a) 5.1 GHz, (b) 5.5 GHz, and (c) 5.9 GHz.

2.2, the axial ratio in the vertical direction with the slot pairs will be the best and the value will be the minimum, while in the other nonvertical direction with the slot pairs the axial ratio will deteriorate. Because four slot pairs are arranged around the axis of coaxial cylinder in the out conductor, the slot pairs are noncoplanar. When rotated around the axis of the antenna, the axial ratio pattern in the xy-plane will be small rippled almost regularly as the symmetrical antenna structure in different ϕ directions.

The measured gain and average axial ratio in the omnidirectional plane results are shown in **Figure 14**. It can be seen that the gain is from 4 to 6 dBic within 5.0–6.0 GHz, and the average axial ratio in xy-plane is below 1.5 dB within the bandwidth. **Figure 15** shows the simulated total efficiency of the proposed antenna. It can be noted from the efficiency results that the total efficiency of the antenna from 5.1 to 5.9 GHz is from 86.9 to 93.5%, so the antenna works with very high efficiency within its operation bandwidth.

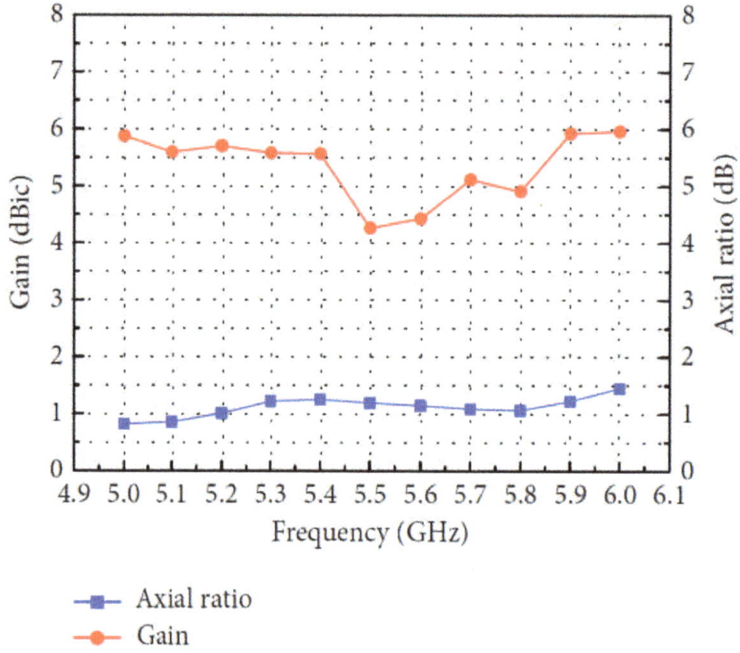

Figure 14. Measured gain and average axial ratio in the omnidirectional plane.

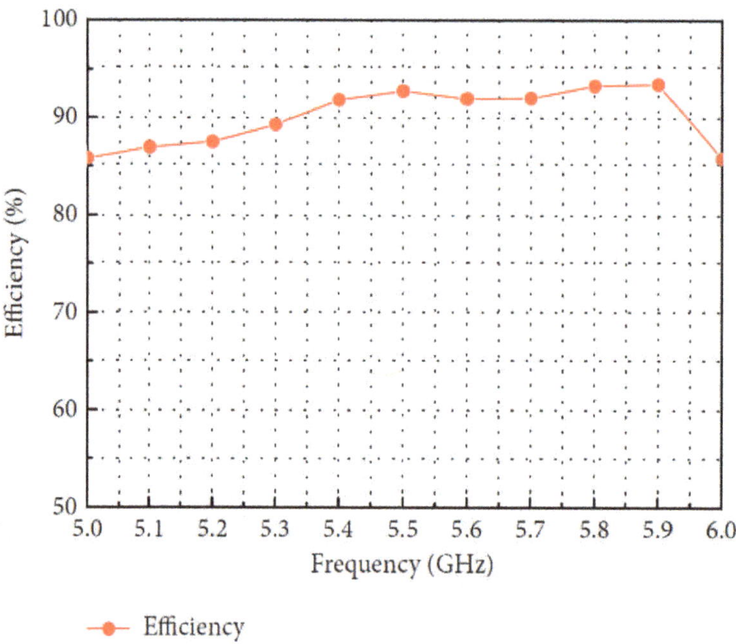

Figure 15. The total efficiency results of the proposed antenna.

4. Compare with others

Here, we compare the proposed omnidirectional CP antenna with others' work in **Table 3**. Our proposed antenna impedance bandwidth, the CP bandwidth is the widest and the gain is largest, although it is the longest.

5. Conclusion

Antenna type	Ref.	Size	Band width	3 dB AR band width	Gain
Omnidirectional CP antenna	[19]	3.175 × 56 × 56 mm³	0.539% (1.481–1.489 GHz)	AR = 2.03 dB (1.481–1.489 GHz)	LHCP: -0.73 to -0.14 dBi RHCP: -21.02 to -17.66 dBi
Dual band omnidirectional CP antenna	[2]	3.175 × 33 × 33 mm³	0.5% (low freq) 1.1% (high freq)	1.57 dB/low freq 0.86 dB/high freq	Low freq: -0.59 to 1 dBi High freq: -0.94 to -0.2 dBi
Omnidirectional CP antenna	[8]	$0.22\lambda_0 \times 0.22\lambda_0 \times 0.07\lambda_0$	3.56%	3.56%	1.5 dBi
Omnidirectional CP antenna	[7, 20]	15 × 90 × 90 mm³	14%	RHCP: 1.35–1.6 GHz LHCP: 1.35–1.65 GHz	RHCP: 0.8 dBi LHCP: 0.1 dBi
DRA omnidirectional CP antenna	[5, 6]	Small	7%	12.3% (2.21–2.5 GHz)	<2 dBi
omnidirectional dual CP antenna	Proposed	190 × 30 × 30 mm³	16.4% (5.05–5.95 GHz)	14.5% (5.1–5.9 GHz)	7 dBi

Table 3. Compare the proposed antenna with others' research results.

The omnidirectional CP antenna with slot array on coaxial cylinder with stable high gain in a wide bandwidth was studied, and the results between measurement and simulation are almost congruent. The improved omnidirectional dual CP antenna is presented too, which achieves the omnidirectional dual CP property by arranging perpendicular slot pairs around and along the coaxial cylinder axis in the out conductor. These antennas have perfectly symmetric structure, so they are easy to change the working frequency band by changing the size of the antenna. For the omnidirectional coverage property and dual CP property, the antennas are valuable in the RF receiver.

Acknowledgements

This work was supported by the National Natural Science Foundation under Grants 61471240,61571289, and 61571298 and the Project of "SMC Excellent Young Faculty."

Author details

Bin Zhou, Junping Geng*, Xianling Liang, Ronghong Jin and Guanshen Chenhu

*Address all correspondence to: gengjunp@sjtu.edu.cn

Department of Electric Engineering, Shanghai Jiao Tong University, Shanghai, China

References

[1] K. Sakaguchi and N. Hasebe, "A circularly polarized omnidirectional antenna," in Proc. 8th ICAP, 1993, pp. 477–480.

[2] B. C. Park and J. H. Lee, "Dual-band omnidirectional circularly polarized antenna using zeroth- and first-order modes," IEEE Antennas Wireless Propag. Lett., vol. 11, pp. 407–410, 2012.

[3] A. Narbudowicz, X. L. Bao, and M. J. Ammann, "Dual-band omnidirectional circularly polarized antenna," IEEE Trans. Antennas Propag., vol. 61, no. 1, pp. 77–83, 2013.

[4] W. Q. Cao, A. J. Liu, and B. N. Zhang, "Dual-band spiral patch-slot antenna with omnidirectional CP and unidirectional CP properties," IEEE Trans. Antennas Propag., vol. 61, no. 4, pp. 2286–2289, 2013.

[5] Y. M. Pan, K. W. Leung, and K. Lu, "Omnidirectional linearly and circularly polarized rectangular dielectric resonator antennas," IEEE Trans. Antennas Propag., vol. 60, no. 2, pp. 751–759, 2012.

[6] W. W. Li and K. W. Leung, "Omnidirectional circularly polarized dielectric resonator antenna with top-loaded alford loop for pattern diversity design," IEEE Trans. Antennas Propag., vol. 61, no. 8, pp. 4246–4256, 2013.

[7] B. Li, S. Liao, and Q. Xue, "Omnidirectional circularly polarized antenna combining monopole and loop radiators," IEEE Antennas Wireless Propag. Lett., vol. 12, pp. 607–610, 2013.

[8] Y. Yu, Z. Shen, and S. He, "Compact omnidirectional antenna of circular polarization," IEEE Antennas Wireless Propag. Lett., vol. 11, pp. 1466–1469, 2012.

[9] X. Quan, R. Li, and M. M. Tentzeris, "A broadband omnidirectional circularly polarized antenna," IEEE Trans. Antennas Propag., vol. 61, no. 5, pp. 2363–2370, 2013.

[10] K. Iigusa, T. Teshirogi, M. Fujita, S.-I. Yamamoto, and T. Ikegami, "A slot-array antenna on a coaxial cylinder with a circularly polarized conical beam," Electron. Commun. Jpn. Pt. I, vol. 83, pp. 74–87, 2000.

[11] W. Croswell and C. Cockrell, "An omnidirectional microwave antenna for use on spacecraft," IEEE Trans. Antennas Propag., vol. 17, no. 4, pp. 459–466, 1969.

[12] C. Y. Yu, T. H. Xu, and C. J. Liu, "Design of a novel UWB omnidirectional antenna using particle swarm optimization," Intl. J. Antennas Propag., vol. 2015, Article ID 303195, 7 p., 2015.

[13] M. Khalily, M. R. Kamarudin, M. Mokayef, and M.H. Jamaluddin, "Omnidirectional circularly polarized dielectric resonator antenna for 5.2-GHz WLAN applications," IEEE Antennas Wireless Propag. Lett., vol. 13, pp. 443–446, 2014.

[14] X. D. Bai, X. L. Liang, M. Z. Li, B. Zhou, J. Geng, and R. Jin, "Dual-circularly polarized conical-beam microstrip antenna," IEEE Antennas Wireless Propag. Lett., vol. 14, pp. 482–485, 2015.

[15] D. Yu, S.-X.Gong, Y.-T.Wan, and W.-F.Chen, "Omnidirectional dual-band dual circularly polarized microstrip antenna Using TM01 and TM02 modes," IEEE Antennas Wireless Propag. Lett., vol. 13, pp. 1104–1107, 2014.

[16] X.-L. Quan and R.-L. Li, "Broadband dual-polarized omnidirectional antennas," in Proceedings of the IEEE Antennas and Propagation Society International Symposium (APSURSI '12), pp. 1–2, IEEE, Chicago, Ill, USA, July 2012.

[17] B. Zhou, J. P.Geng, X. D. Bai, L. Duan,X. Liang, and R. Jin, "An omnidirectional circularly polarized slot array antenna with high gain in a wide bandwidth," IEEE Antennas Wireless Propag. Lett., vol. 14, pp. 666–669, 2015.

[18] M. E. Bialkowski and P. W. Davis, "Linearly polarized radialline slot-array antenna with a broadened beam," Microwave Optical Technol. Lett., vol. 27, no. 2, pp. 98–101, 2000.

[19] B.C. Park and J.H. Lee, "Omnidirectional circularly polarized antenna utilizing zeroth-order resonance of epsilon negative transmission line," IEEE Trans. Antennas Propag. vol. 59, no. 7, pp. 2717–2721, 2011.

[20] B. Li and Q. Xue, "Polarization-reconfigurable omnidirectional antenna combining dipole and loop radiators," IEEE Antennas Wireless Propag. Lett., vol. 12, pp. 1102–1105, 2013.

4

Investigating EM Dipole Radiating Element for Dual-Polarized Phased Array Weather Radars

Ridhwan Khalid Mirza, Yan (Rockee) Zhang,
Dusan Zrnic and Richard Doviak

Abstract

Dual-polarized antenna radiating element is a critical component in the Multi-function Phased Array Radar (MPAR). This paper studies the dual-polarized radiating element based on the EM dipole concept. Two different geometries, i.e., loop approximated as magnetic dipole and a printed electric dipole, are used to form a single dual-polarized radiating element. Radiation patterns based on Ansoft HFSS (High Frequency Structural Simulator) simulation software and measurements carried out in anechoic chambers are presented. Initial array design based on these elements will also be discussed.

Keywords: MPAR, dual-polarized elements, cross pol, E&M dipole array, loop antenna, electric dipole

1. Introduction

Multi-function Phased Array Radars (MPARs) are being considered for fulfilling the need of FAA, NOAA/NWS, and possibly Homeland security [1]. One of the specific goals of deploying MPAR systems is to replace the current Weather Surveillance Radar (WSR-88D radar) and other air-traffic control radars as upgrade [2]. In many radar applications and weather radar polarimetry, the physical symmetry of the target and weather scan moments can be detected by phased array antennas which require suppression of antenna cross polarization (X-pol) especially during the main beam shift or scanning off-broadside direction [3, 4]. Dual-polarized antenna radiating element plays a key role in such systems. The current state-of-the-art parabolic reflector antennas which are used for weather surveillance are required to scan mechanically and lack the ability to scan the beam electronically in contrast to phased array radars. The cross polarization (X-pol) isolation of phased array radar can be improved effectively if the cross polarization levels of individual antenna radiating elements are minimized.

Orthogonally signals can be transmitted or received without extra bandwidth requirement or physical separation between antennas through a dual-polarized radiating element, which enables observing backscattering from targets (precipitation, clutter, etc.) from horizontally and vertically polarized electromagnetic waves. For instance, vertical tree trunks and structures can be tracked by HH polarization. VV polarization can provide water and soil scattering information [5]. However, for accurate polar metric measurements, it is required for the antenna to have high cross polarization suppression. The cross polarization isolation as required by many weather polarimetry applications is less than 20 dB for alternate transmission and less than 40 dB for simultaneous transmission and reception [2, 6].

There are many other forms of antennas in the form of microstrip patches which are used widely in Multi-Functional Phased Array Radars (MPARs) [4, 7, 8]. Although better than -30 dB cross pol levels are achieved in most of the rectangular patch-based MPAR subarray prototypes, however, there are certain challenges which arise when steered-beam calibrations are needed for precise dual-pol measurements. The main challenge arises when scanning the beam off-broadside direction, and the relationship between co-pol and cross pol electric fields is not consistent for different pointing angles of an antenna [8]. An ideal radiating element should be able to radiate electric fields which are orthogonal to each other for two different polarizations [2, 9]. To achieve this, two types of dual polarization elements are considered. The first one is a pair of cross dipole or orthogonal dipole element configuration in which the electric fields are orthogonal only in principal planes which is under investigation by many researchers [10–12]. The other type of radiator is a pair of magnetic dipole and electric dipole for which the electric fields are orthogonal in all spatial directions. The latter configuration is investigated and proposed in this book chapter as the dual-polarized radiating element.

If these types of elements which have orthogonal electric fields are used, especially for cylindrical phased array radars (CPARs), the problems of surface and traveling waves can be effectively avoided [8]. Presently, the main disadvantages of using the dipole elements are higher costs for fabrication and potential requirements of three-dimensional structures. However, these disadvantages outweigh the potential benefits of using such a configuration especially for polarimetric phased array radars which have strict constraint of maintaining orthogonality between two different polarizations.

This chapter/proposal studies different antenna geometries such as loop and planar dipoles. The characteristics of these radiating element designs are evaluated with HFSS simulations. The simulations show that these structures result in unidirectional circular current distribution; improvement at all aspects of the antenna characteristics including radiation pattern, return loss, and cross polarization isolations at single element level will be discussed. The far-field radiation patterns for the radiating elements will be displayed as measured in anechoic chambers at Radar Innovations Lab, University of Oklahoma.

2. Background

This section describes the basic theory on the electromagnetic fields, antenna coordinate system, and basic definitions used in this chapter.

2.1. Electromagnetic field regions around an antenna

The antenna radiates waves in the form of electromagnetic radiation. Hence, defining the field regions around an antenna is lucrative to know, and this will be used for the later in the chapter while discussing results and radiation patterns. The space surrounding any antenna is divided into three regions based on the distance from the antenna under test (AUT). These regions are based on equations as described **Table 1** [9].

Table 1 shows three different regions around an antenna based on the distance. These are calculated based on the equations where r is the distance from the source or AUT, D is the maximum dimension of the AUT, and λ is the wavelength of the radiation. The field region of antenna will be helpful to study and analyze the antenna radiation patterns discussed later in the chapter. Most of the times, the targets are placed or objects are measured in the far-field or Fraunhofer region. These regions are also illustrated in **Figure 1**.

2.2. Antenna coordinate system

The radiation characteristics of an antenna are analyzed using a coordinate system. There are different types like cylindrical, planar, and spherical. However since the antenna for the phased array is to be used in the far field, the spherical coordinate system is optimal as shown in **Figure 2**. In this system the antenna is placed at the origin and r is the distance or radius from the AUT to the observation point. The radiation characteristics of an AUT are determined by varying the elevation or zenith angle θ and azimuth angle ϕ, and field vectors E_θ, E_ϕ are calculated. This spherical coordinate system is used in this chapter for simulation and carrying out measurements in the anechoic chambers.

Region	Distance
Near-field or reactive near-field region	$0 < r < 0.62\sqrt{\frac{D^3}{\lambda}}$
Reactive near-field or Fresnel region	$0.62\sqrt{\frac{D^3}{\lambda}} \le r \le 2\frac{D^2}{\lambda}$
Far-field or Fraunhofer region	$2\frac{D^2}{\lambda} \le r$

Table 1. Field regions around an antenna.

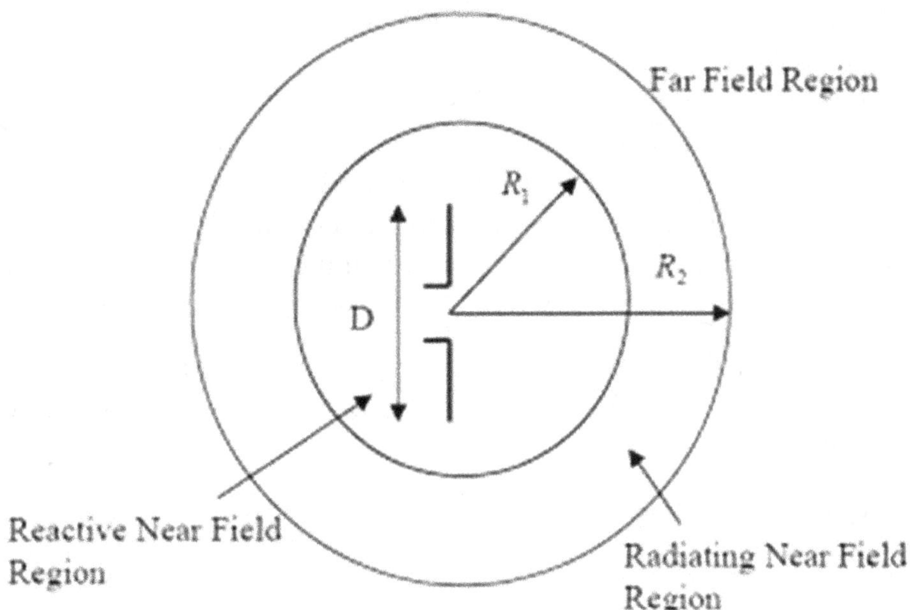

Figure 1. Field regions around an antenna.

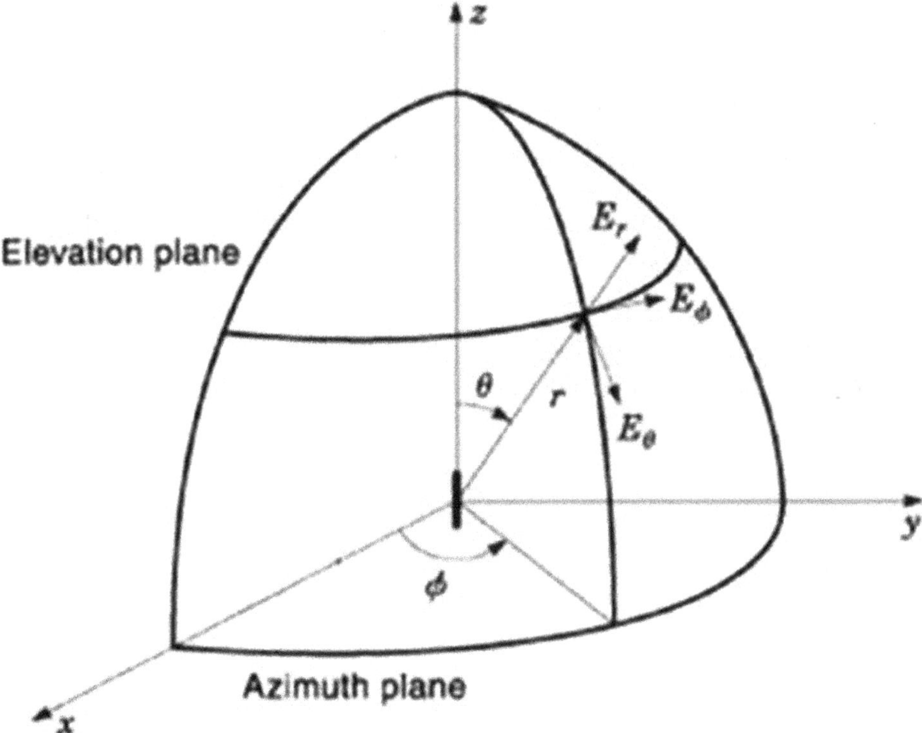

Figure 2. Spherical coordinate system used for antenna analysis [9].

2.3. Antenna polarizations

The polarization of the radiated electric fields produced by an antenna in a given direction is referred as polarization of an antenna. The polarizations of antenna depend on the way the antenna radiates waves and can be studied in open literature [9].

In weather radar polarimetry, there are two important polarization definitions: copolarization (often called as co-pol) and cross polarization (often called as cross pol or X-pol) of a radiating element. The term copolarization is the desired polarization component of the radiation pattern, and cross polarization is the unwanted component of radiation pattern. For instance, if the transmitting antenna is meant to transmit horizontally polarized waves and the receiving antenna on the other end is receiving horizontally polarized waves, it is said to receive co-pol radiation or desired radiation. On the contrary if the receiving antenna was supposed to receive vertically polarized waves, it is receiving cross polarization or unwanted radiation.

There are three most widely used definitions for cross pol and co-pol as given by Ludwig [13]. For weather radars, the second definition of Ludwig (L2) is widely accepted to define the co- and cross polarizations (**Figure 3**).

According to Ludwig's second definition [13], we have

$$\hat{\imath}_{ref} = \frac{\sin\phi\cos\theta\ \hat{\imath}_{\theta} + \cos\phi\,\hat{\imath}_{\phi}}{\sqrt{1-\sin^2\theta\sin^2\phi}} \tag{1}$$

$$\hat{\imath}_{cross} = \frac{\cos\phi\,\hat{\imath}_{\theta} - \cos\theta\sin\phi\,\hat{\imath}_{\phi}}{\sqrt{1-\sin^2\theta\sin^2\phi}} \tag{2}$$

where $\hat{\imath}_{ref}$ and $\hat{\imath}_{cross}$ are the projections of electric field vectors onto spherical unit vectors given by Ludwig [13] such that,

$E.\hat{\imath}_{ref}$ = the reference polarization (co-pol) component of E

$E.\hat{\imath}_{cross}$ = the cross polarization (cross pol) component of E

The spherical coordinate system used for the weather radar polarimetry and most of the standard antenna measurements is as depicted in **Figure 2**, in which the polar axis is along *z*-axis and antenna axis is aligned along *z*-axis. After this the Ludwig's II definition (transformation) is applied to obtain the co-polar and cross polar patterns. Hence, all the measurements shown in this chapter will be as per L2 definition of cross polarization.

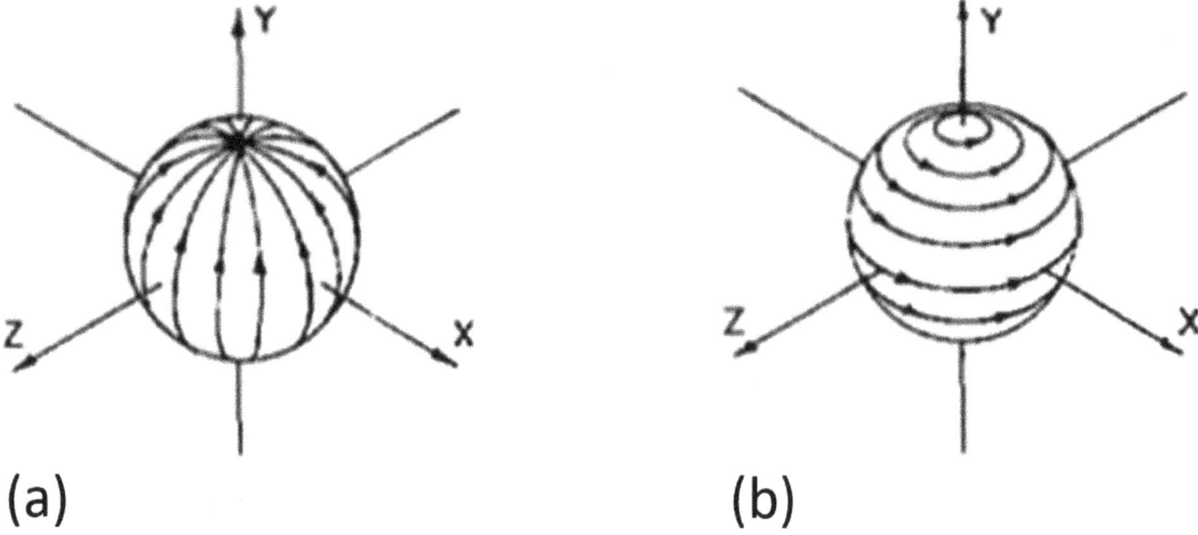

Figure 3. Ludwig II coordinate system [13], (a) direction of reference polarization and (b) direction of cross polarization.

3. Loop as a magnetic dipole

Loop antennas are one of the simple, versatile, and inexpensive forms of microstrip antenna. In this section, the basic condition for a loop to exhibit magnetic dipole characteristics is discussed. The different implementations of loop antennas in order to achieve its requirements are tested. It will be shown that a loop antenna with capacitive loading has uniform current distributions and its simulation results are discussed.

3.1. Magnetic fields due to constant current loop

The loop antenna can behave as magnetic dipole when it has constant current along its circumference. To illustrate this, consider a loop with constant current i, radius r, and an observation point at a far-field distance R as shown in **Figure 4**. At far-field distance where ($R \gg r$), the loop can be treated as a small circular loop with constant current, and the calculation of magnetic field depends on the current i, R, and θ, i.e., the angle form the z-axis. The magnetic field equations are given by [14]

$$B = \begin{cases} B_r = 2|\mu|\mu_0 \dfrac{cos\theta}{4\pi R^3} \\ B_\theta = |\mu|\mu_0 \dfrac{sin\theta}{4\pi R^3} \end{cases} \tag{3}$$

Where $\mu = iA$ is the magnetic dipole moment of the loop and A is the area of the loop. These magnetic fields produced by the loop are equivalent to the fields produced by a small magnetic dipole. The magnetic dipole moment as stated above is a vector pointing out normal to the plane of the loop, and its magnitude is equal to the product of current and area of the loop.

3.2. Design and simulation results of loop based on capacitive loading

As aforementioned in Section 3.1, the loop antenna can have the same characteristics like a magnetic dipole if constant unidirectional current is maintained along its circumference.

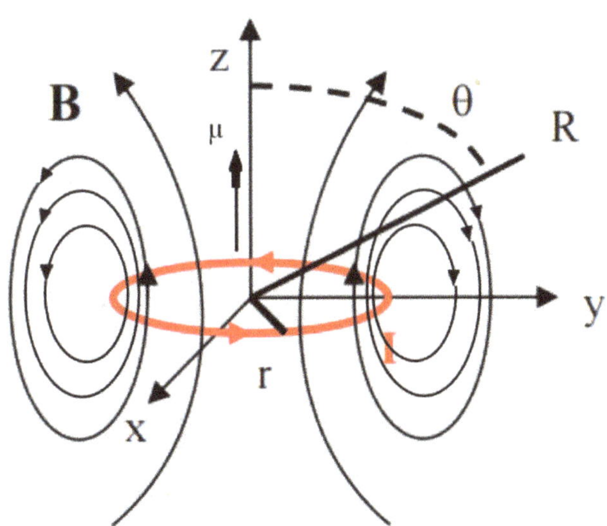

Figure 4. Illustration of magnetic fields around a loop.

This is generally difficult to achieve for smaller loop antennas because of high reactance and small radiation resistance which leads to improper matching [9, 15]. To combat this problem, Foster [16] derived the idea of driving the loop currents in segments which also explains the analytical solution for radiation resistance and directivity for these types of loops. Later Li et al. [17] proved that by adding a capacitive reactance on the loop at every 45°, a uniform traveling wave current distribution can be obtained. A similar concept was used by Wei et al. [18] for designing a horizontally polarized loop antenna. The loop antenna shown in **Figure 5** is reproduced using the original design by Wei et al. [18] with different substrate material and different operating frequencies.

This proposed loop shown in **Figure 5** has periodic capacitive loading at every 45° and consists of strip-line sections which are similar to unit cell described by Park et al. [19]. This unit cell which is periodically placed helps to maintain uniform in-phase surface current distribution on the surface of the loop as seen in **Figure 5(b)** by creating a series capacitance. Therefore, this antenna can meet the condition of behaving as a magnetic dipole in far field.

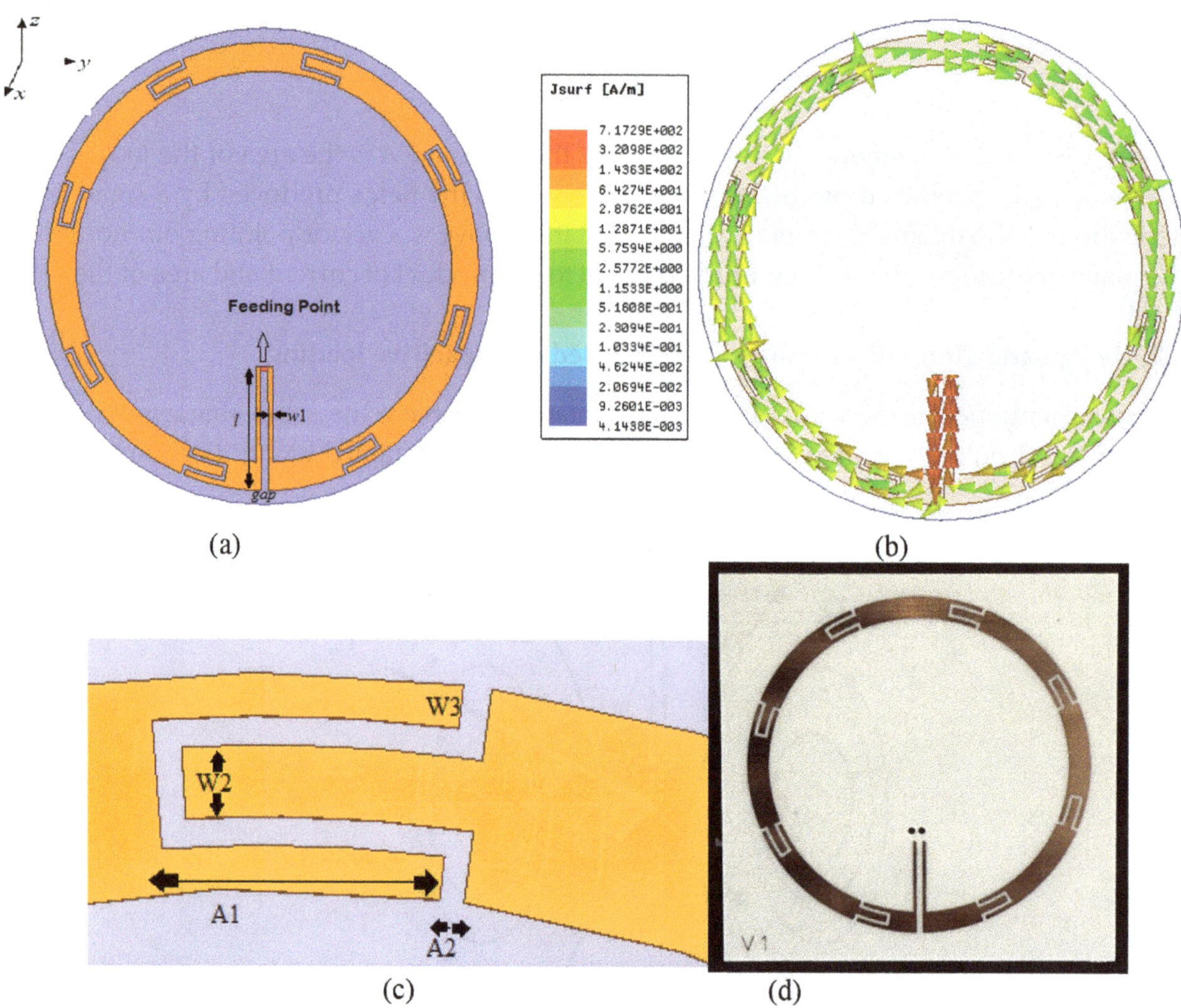

Figure 5. Loop with capacitive loading: (a) design in HFSS, (b) illustration of surface current, (c) dimensions of the unit cell, and (d) fabricated prototype.

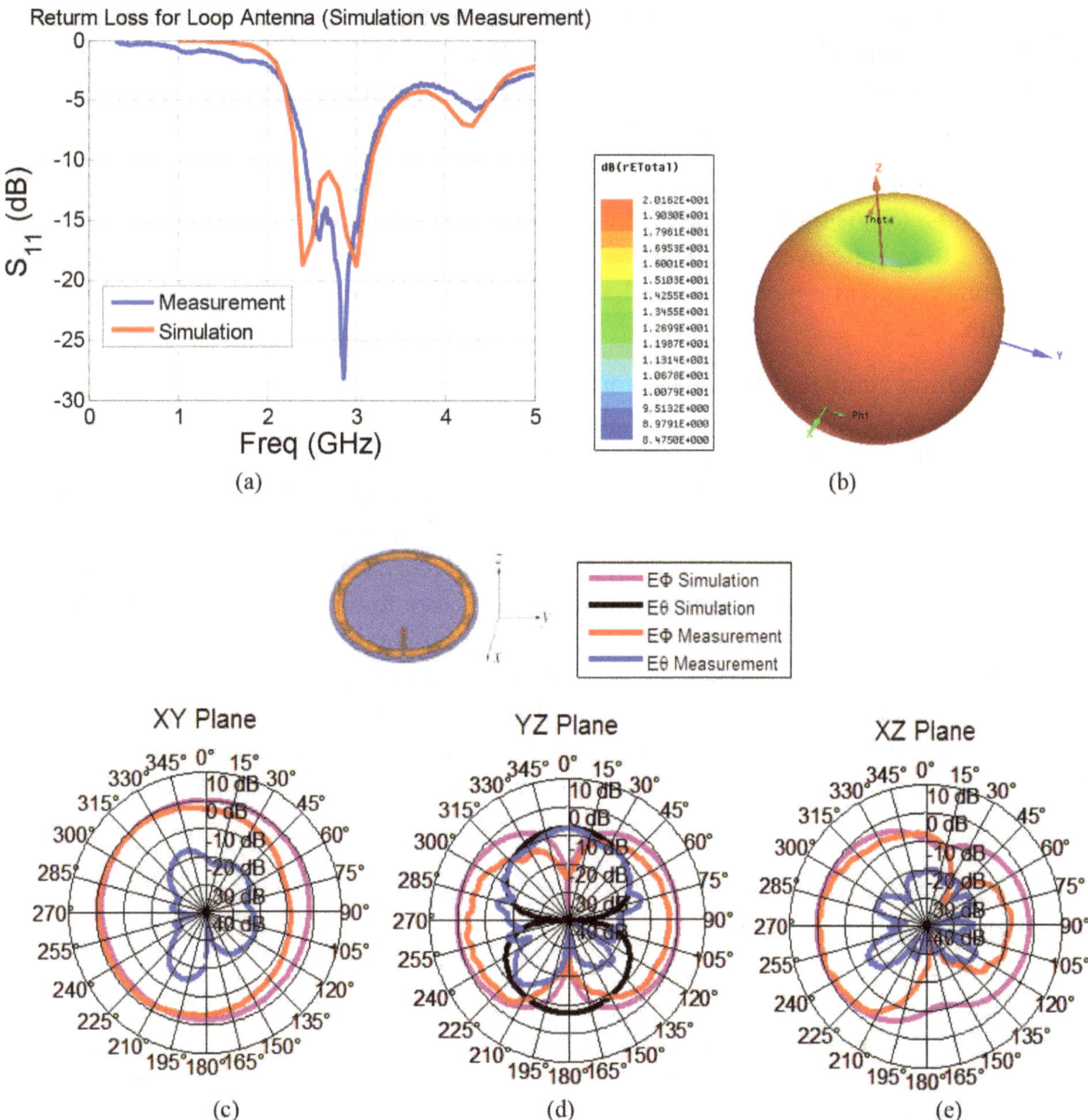

(a) (b)

(c) (d) (e)

Figure 6. Comparison of measured and simulated pattern: (a) return loss (S_{11}) in dB, (b) total 3D radiation pattern, (c) XY plane cut of the radiation pattern, (d) YZ plane cut of the radiation pattern, and (e) XZ plane cut of the radiation pattern.

The loop is designed on Rogers 4725JXR substrate ($\varepsilon_r = 2.64$) with 0.787 mm thickness with 1 oz. copper. The loop has inner radius (R2) = 20.5 mm and outer radius (R1) = 23.5 mm. Each unit cell of strip line has the following dimensions as depicted in **Figure 5(c)** as $A_1 = 11°$ or 4 mm, $A_2 = 1°$ or 0.35 mm, $w_2 = 1$ mm, and $w_3 = 0.6$ mm. The periodic unit strip-line sections (unit cells) are placed at every 45°. The number of unit cells for capacitive loading are related by relation $A_1 + A_2 = \frac{360}{N}$, where N being the number of unit cells. Another advantage of using the periodic capacitive loading is to achieve a wide impedance bandwidth [18]. The antenna is fed by using a parallel strip-line which is a balanced structure and also plays a role of a balun and helps in maintaining impedance matching. Each strip line has a length (l) = 13 mm, width (w_1) = 0.5 mm, and gap = 0.8 mm as seen in **Figure 5(a)**. The structure was simulated in Ansoft HFSS simulation software. The absorbing boundary conditions (ABCs) were used in the simulation environment, and the antenna is excited by using lumped port excitation. The ABC boundaries are placed at a distance more than $\frac{\lambda}{4}$ from the antenna.

The SMP connector at the feeding point for the fabricated prototype is shown in **Figure 5(d)**. The return loss (S_{11}) is measured using the PNA and can be seen in **Figure 6(a)**. The antenna is designed to operate over 2.3–3.1 GHz in simulation. However the measured results show that antenna has good return loss over 2.5–3.0 GHz which gives us at least about 500 MHz of bandwidth.

The antenna was simulated at 2.8 GHz and the simulation and measurement results for principal planes are shown in **Figure 6**. As the loop is horizontally polarized, it radiates horizontally polarized fields (E_ϕ, i.e., co-polar radiation for this case). There are also vertically polarized fields (E_θ, i.e., the cross polar radiation for this case) which are unwanted electric fields (orthogonal to desired direction) from the loop. These vertically polarized fields usually exist in the real world due to feeding point variations and material imperfections. It can be observed that the radiation fields of the loop are close to an ideal magnetic dipole as it exhibits a pattern similar to that of a doughnut shape with a null along its axis (in elevation plane) and has maximum radiation along the plane of the loop (XY plane). Ideally, the radiation in XZ plane and YZ plane should be similar for a magnetic dipole; however, because of the feed line design of this loop, there is some electric field radiation from this feed line (two parallel strip lines). This adds up to the cross polar fields and constitutes stronger E_θ in YZ plane.

4. Printed electric dipole

An electric dipole antenna is widely used for many applications and is considered as one of the simplest form of linear wire antennas [20]. The design of a regular linear wire dipole is simple. It consists of a conductor (wire) which is split in the middle to allow the feeder to transmit or receive power.

As shown in **Figure 7**, a dipole consists of two poles; the length of each pole is quarter wavelength ($\frac{\lambda}{4}$), thereby making overall dipole as half-wavelength ($\frac{\lambda}{2}$). The dipole radiates equal power in all azimuthal directions perpendicular to the axis of the antenna. Section 4.1 describes the design and simulation of the dipole used in for this research.

4.1. Design and simulation results of a printed electric dipole

Several types of dipoles are available such as wire dipole, folded dipole, printed dipole, cage dipole, and bow tie and batwing antenna [20]. These are specific to the type of application and decisions of design. In this chapter, the printed dipole is chosen among different variations of dipole antennas, because it is easy to design and does not involve 3D structures. The advantage of light weight, small size, and planar form of dipole can be realized when it is aligned with loop which will be discussed in the next section.

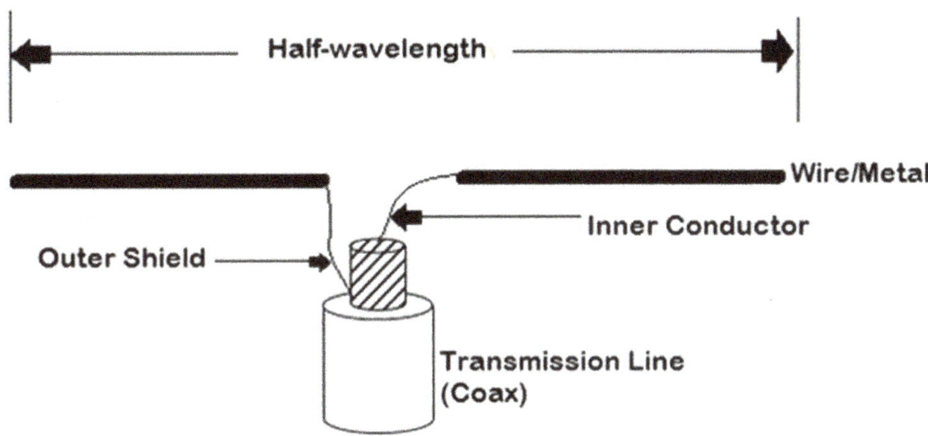

Figure 7. A simple linear wire dipole antenna.

The proposed design is a very simple planar printed electric dipole. This microstrip dipole antenna has two arms or poles as shown in **Figure 8**. Each of the arms has length *L* and width *W* and is separated by a gap *Gap*.

The microstrip printed dipole is designed on Rogers 4725JXR ($\varepsilon_r = 2.64$) with 0.787 mm thickness and 1 oz. copper. The design is based on the equations given below [21]:

$$\varepsilon_e = \frac{\varepsilon_r + 1}{2} + \frac{\varepsilon_r - 1}{2}\left[\frac{1}{\sqrt{1 + 12d/W}}\right] \tag{4}$$

where ε_r = dielectric constant; ε_e = effective dielectric constant; d = substrate thickness; W = width of microstrip line.

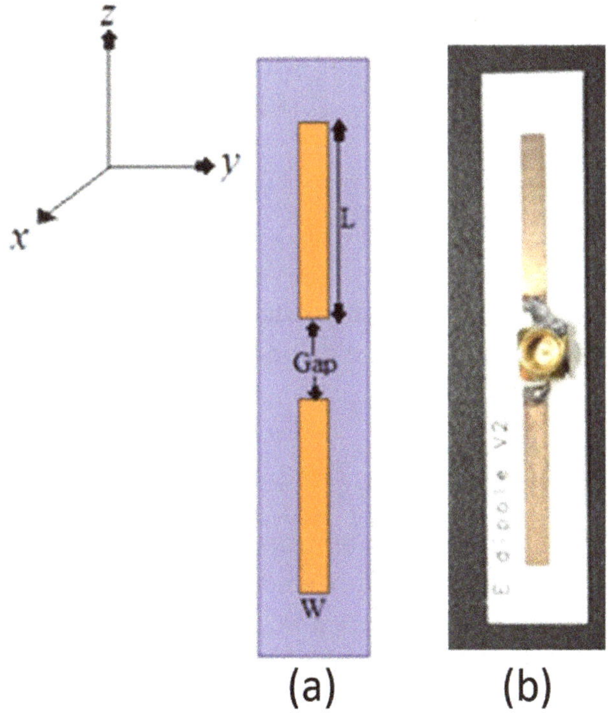

Figure 8. A printed dipole: (a) design and HFSS model and (b) fabricated prototype.

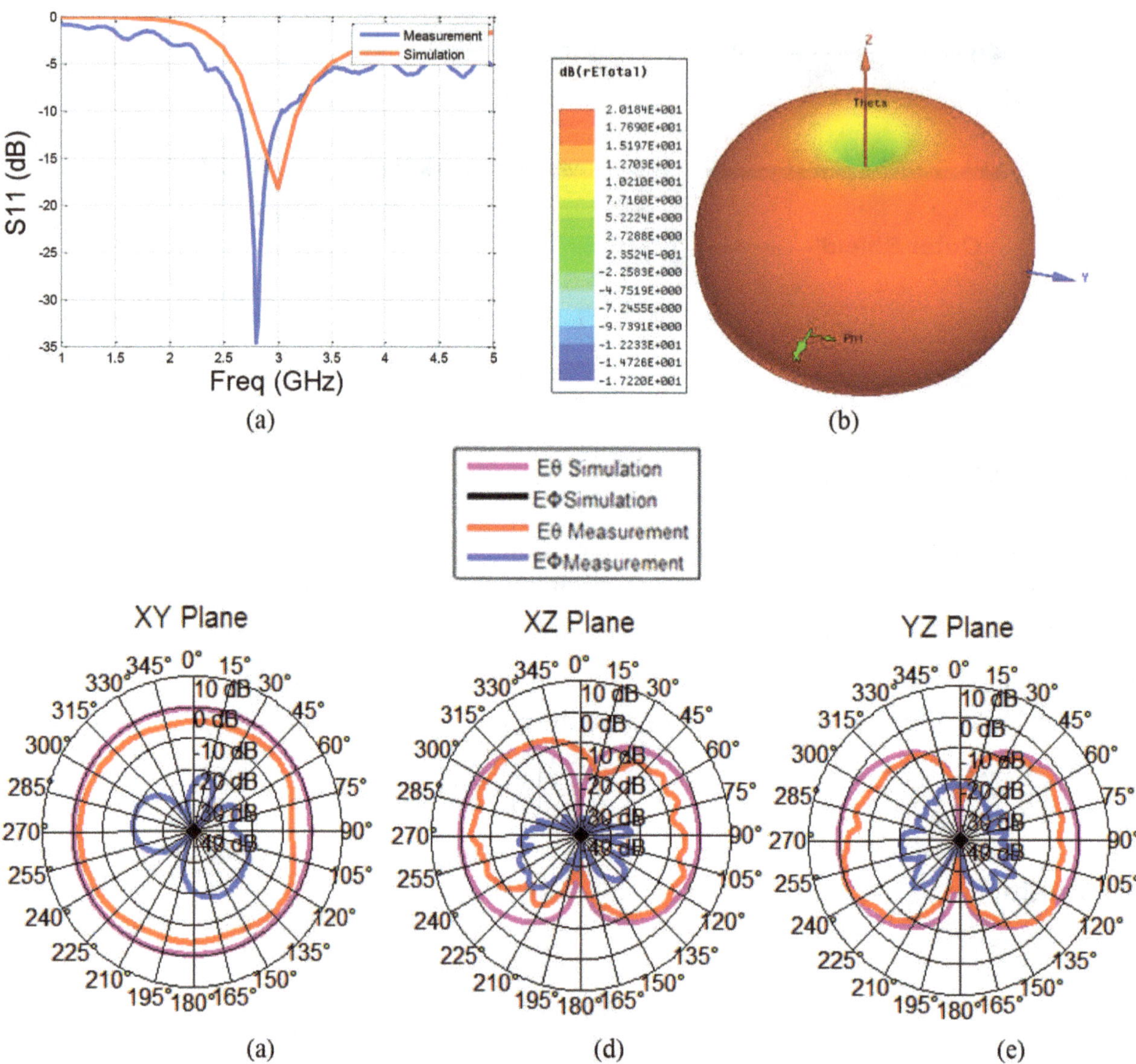

Figure 9. HFSS simulated return loss and radiation patterns of the proposed electric dipole antenna: (a) return loss (S_{11}) in dB, (b) total 3D radiation pattern, (c) XY plane cut of the radiation pattern, (d) XZ plane cut of radiation pattern, and (e) YZ plane cut of radiation pattern..

The length *L* of the printed dipole is approximated to quarter wavelength ($\frac{\lambda}{4}$) and the entire dipole constitutes half wavelengths ($\frac{\lambda}{2}$). The value "λ" is given by

$$\lambda = \frac{c}{f\sqrt{\varepsilon_e}} \tag{5}$$

where λ = wavelength; c = velocity of light; f = frequency.

These values are tuned and approximated by using the parametric setup in HFSS simulation software. The values used for the above fabricated prototype are L = 17 mm, W = 2.5 mm, and *Gap* = 7 mm. The antenna is excited by lumped port excitation in HFSS and absorbing boundary conditions (ABCs) was used in the simulation environment. The ABC boundaries are placed at a distance more than $\frac{\lambda}{4}$ from the antenna.

The microstrip electric dipole is operating over 400 MHz bandwidth ranging from 2.7 to 3.1 GHz.

It is seen in **Figure 9(a)** that there is a slight shift in frequency in measured results as compared to simulation. This might be due to the effect of use of SMP connector as shown in **Figure 8(b)** practically, versus an ideal 50 Ω matched lumped port in simulation. The simulation results show the radiation patterns at 2.8 GHz (selected for the convenience of comparing them with measurement results of electric dipole and loop) for principal planes seen in **Figure 9**. Although the electric dipole is designed to radiate vertically polarized fields (E_θ, i.e., the co-polar radiation for this case), it is observed that there are also horizontally polarized fields (E_ϕ, i.e., the cross polar radiation) radiated from the antenna. This cross polar radiation might be due to the factors like antenna aperture, radiation of electric field in unwanted direction (orthogonal to desired direction), material imperfections, and radiation from the feed point. The simulation results show that the plane of maximum radiation is in XY plane, and there is a null along z-axis which is part of the doughnut shape as seen in **Figure 9(b)**.

5. Dual-polarized radiating element

By discussing the design of loop antenna in the previous section, it is clear that the loop antenna has capability to radiated fields which are similar to that of a magnetic dipole. The proposed design of the loop as in Section 3 is an electrically large loop as the circumference of loop is greater than one-tenth of wavelength ($\lambda/10$). But at a far-field distance from the source ($R \gg r$), where R is the observation point and r is the radius of the loop, the loop can be considered as an electrically small loop. This electrically small loop antenna acts as a dual antenna to a short dipole antenna. The far-field electric and magnetic fields of magnetic dipole (loop) are identical to the far-field magnetic and electric field of electric dipole, respectively [22]. This can be illustrated in **Figure 10** [22].

Consider a loop having a uniform current I_0' and area $\Delta S'$. The far-field vector potential for electrically small loop is given by [9, 22] as

$$A \approx j\frac{k\mu I_0 \Delta S}{4\pi r} e^{-jkr} \sin\theta \; a_\phi \qquad (6)$$

where $\Delta S = \Delta l^2$, i.e., the area of the loop and a_ϕ is the spherical coordinated vector given by $a_\phi = (-\sin\phi \, a_x + \cos\phi \, a_y)$.

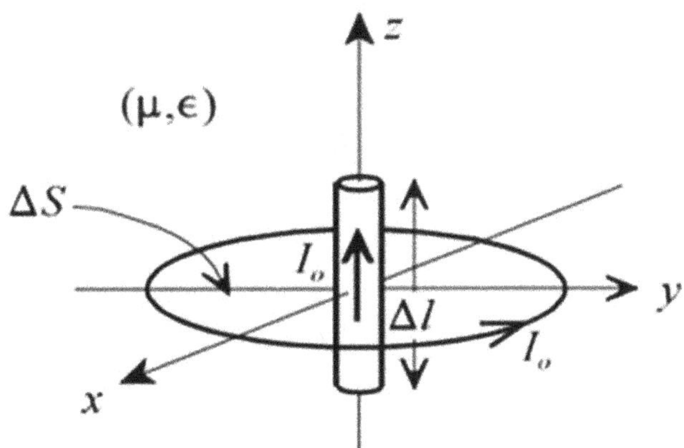

Figure 10. Illustration of collocated loop and electric dipole arrangement [22].

The corresponding far-field components for electrically small loops can be calculated using Eq. (6) as [9, 22]

$$E_\phi \approx -j\omega A_\phi \approx \frac{\eta k^2 I_0 \Delta S}{4\pi r} e^{-jkr} sin\theta \tag{7}$$

$$H_\theta \approx j\frac{\omega}{\eta} A_\phi \approx -\frac{k^2 I_0 \Delta S}{4\pi r} e^{-jkr} sin\theta \tag{8}$$

If carefully observed, these components are similar to the fields those radiated by an infinitesimal electric dipole [9, 22]. **Table 2** shows the comparison between these two antennas:

Infinitesimal electric dipole	Electrically small loop
$E_\theta \approx j\frac{\eta k I_0 \Delta l}{4\pi r} e^{-jkr} sin\theta$	$E_\phi \approx \frac{\eta k^2 I_0 \Delta S}{4\pi r} e^{-jkr} sin\theta$
$H_\phi \approx j\frac{k I_0 \Delta l}{4\pi r} e^{-jkr} sin\theta$	$H_\theta \approx -\frac{k^2 I_0 \Delta S}{4\pi r} e^{-jkr} sin\theta$

Table 2. Comparison of far-field elements generated by small electric dipoles and loops.

These equations in **Table 2** show that they share the same mathematical form, and a pair of equivalent and dual source can be found out by changing the symbols. This is known as duality theorem [9]. By duality theorem we have dual quantities of electric and magnetic current sources as shown in **Table 3**.

Now by applying the above dual quantities, the far-field components of the magnetic sources can be determined and are shown in **Table 4**.

Therefore, if an electrically small magnetic loop and an electric dipole are designed such that

$$I_o \Delta l = j\eta k I_{om} \Delta S = j\omega\mu I_{om} \Delta S \tag{9}$$

then due to this equivalence condition, the far fields radiated by the electrically small magnetic loop and electric dipole are equivalent dual sources. Since these antennas act as dual to each other, the power radiated by both should be the same when currents and dimensions are appropriately designed.

Electric source	Magnetic source
E	H
H	$-E$
I_o	I_{om}
K	k
H	$\frac{1}{\eta}$

Table 3. Dual quantities for electric and magnetic current sources.

Infinitesimal magnetic dipole	Electrically small loop (magnetic)
$H_\theta \approx j\frac{kI_{0m}\Delta l}{\eta 4\pi r}e^{-jkr}\sin\theta$	$H_\phi \approx \frac{k^2 I_{0m}\Delta S}{\eta 4\pi r}e^{-jkr}\sin\theta$
$E_\phi \approx -j\frac{kI_{0m}\Delta l}{\eta 4\pi r}e^{-jkr}\sin\theta$	$E_\theta \approx \frac{k^2 I_{0m}\Delta S}{\eta 4\pi r}e^{-jkr}\sin\theta$

Table 4. Far-field components of corresponding magnetic sources.

5.1. Antenna arrangement and polarization

From the above discussion, it is lucrative to determine the antenna configuration/alignment which will be useful to form a dual-polarized antenna unit. In theory, an electric dipole and electrically small loop (magnetic dipole) are dual antennas to each other; their far fields are orthogonal to each other if these antennas are aligned as shown in **Figure 11**. A combination of E&M dipoles is oriented such that the electric field of electric dipole (E_θ) is orthogonal to the electric field of the loop (E_ϕ) everywhere. In other words it can be said as electric dipole should be vertically polarized and loop should be horizontally polarized as shown in **Figure 11(a)**. This can also be illustrated by considering a sphere consisting of latitudes and longitudes. The directions of the fields generated by a loop (magnetic dipole) can be approximated by latitude lines, whereas the direction of the fields generated by electric dipole can be approximated by longitude lines (**Figure 11(b)**). The plane of maximum radiation for the loop and the electric dipole is XY plane. The electric dipole generates quasi vertically polarized waves (i.e., E_θ), and the horizontally polarized loop generates quasi horizontally polarized waves (i.e., E_ϕ).

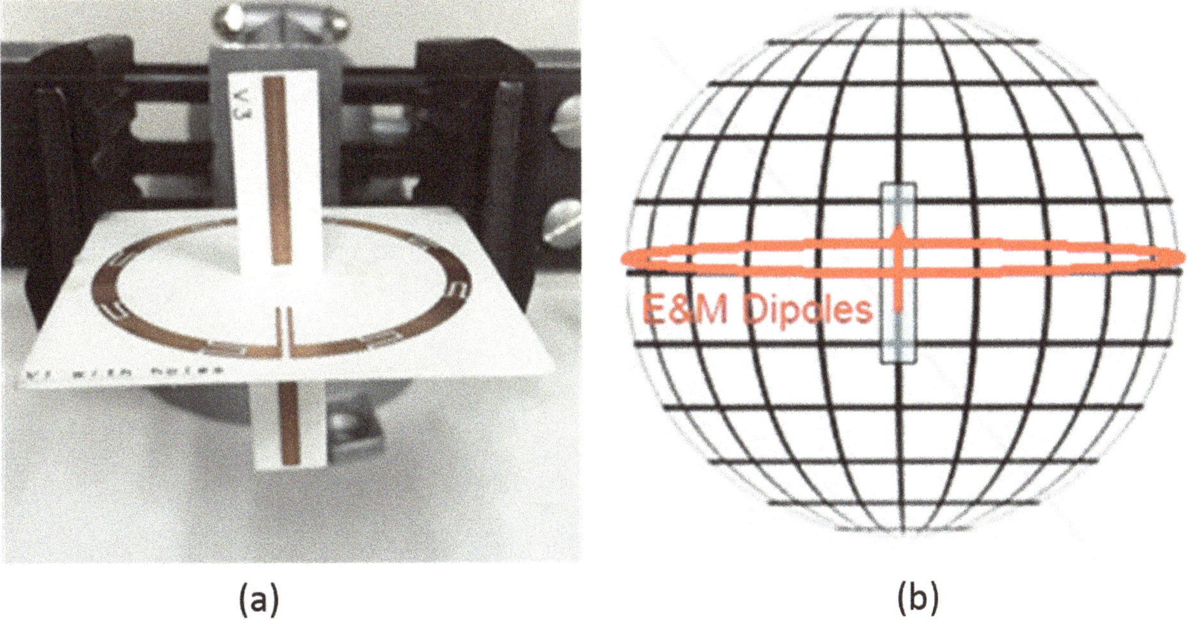

Figure 11. E&M dipole configuration: (a) collocated arrangement and (b) electric field lines due to collinear arrangement of E&M dipoles.

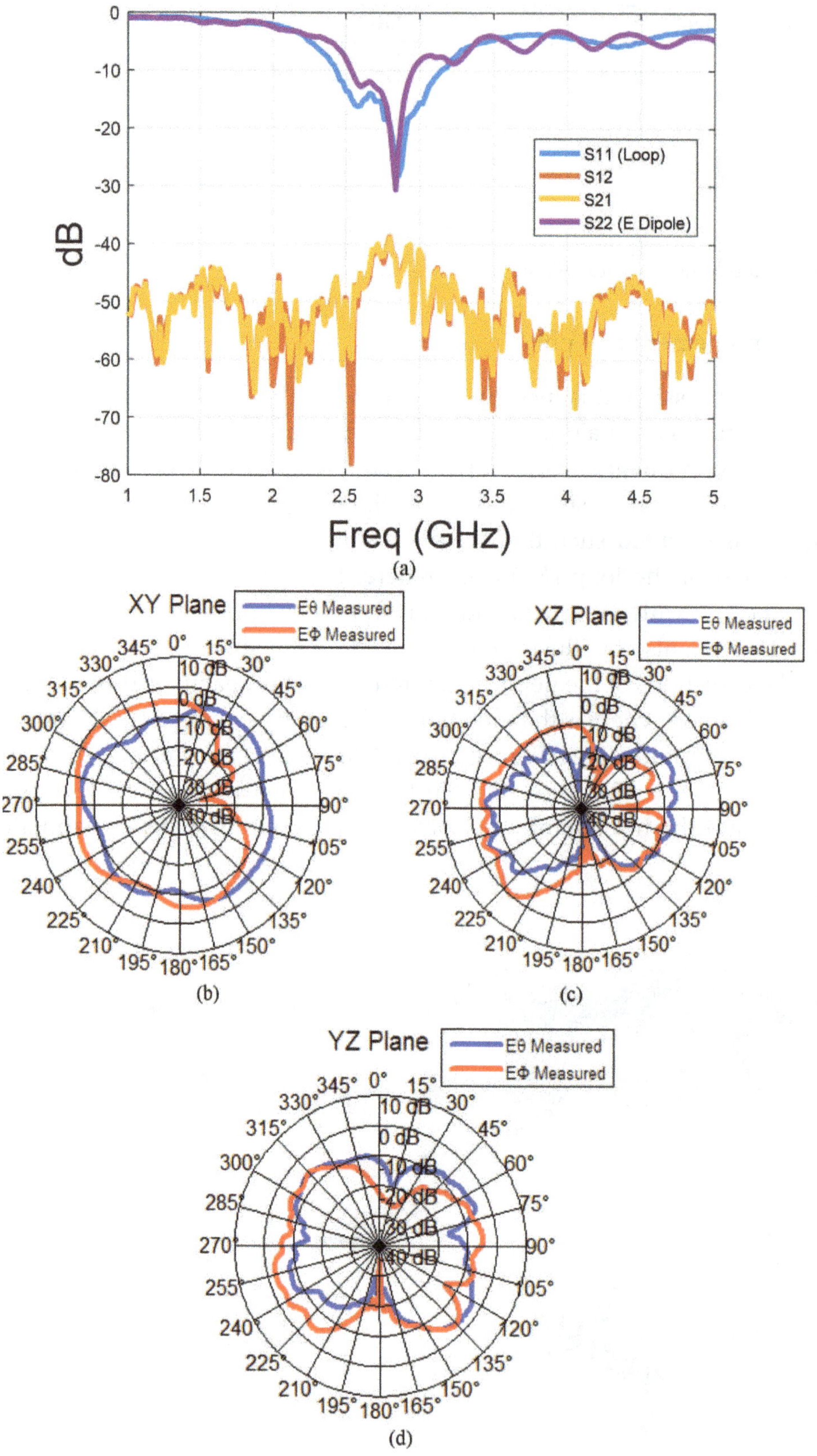

Figure 12. Measurement results when the E dipole antenna and loop antenna are excited simultaneously: (a) measured S-parameters (dB), (b) XY plane cut of the radiation pattern, (c) XZ plane cut of the radiation pattern, and (d) YZ plane cut of the radiation pattern.

This arrangement of electric dipole and loop antenna as discussed above is studied in HFSS simulation, and a fabricated prototype is built as seen in **Figure 11(a)** to measure them in anechoic chambers. The measured S-parameters seen in **Figure 12(a)** show that both electric and magnetic dipoles resonate at 2.8 GHz and have good isolation of about 40 dB in the impedance bandwidth. The measured radiation pattern is illustrated in **Figure 12** for principal planes when both E&M dipoles are excited simultaneously (i.e., H and V simultaneous transmission). The results show that the radiation patterns of both the antennas match very well and these are dual antenna to themselves. The loop generates horizontally polarized waves, and electric dipole generates vertically polarized waves which are orthogonal to each other, thereby serving to be a good candidate for phased array antenna element.

6. E&M dipole array

An array can be built based on the proposed collocated E&M dipole units to emulate a linear subarray for future Multi-functional Phased Array Radar (MPAR). An array can be formed by placing each E&M dipole unit about half wavelengths ($\frac{\lambda}{2}$) from center to center. An investigation of using these elements in a linear form of array is being carried out. The following **Figure 13** shows a simple arrangement of such eight collocated E&M dipole units.

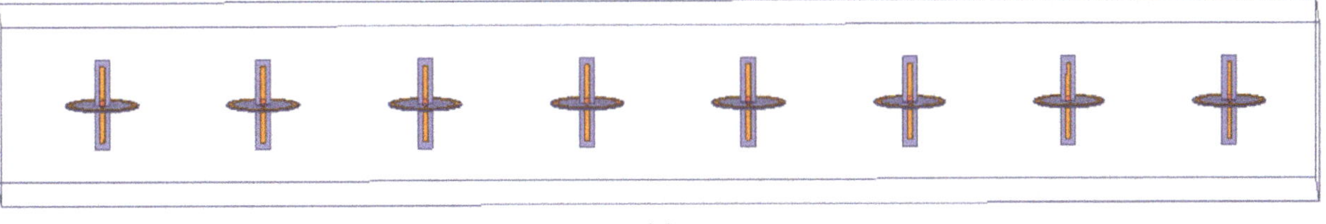

(a)

(b)

Figure 13. An eight-element E&M dipole array: (a) simulation using HFSS and (b) fabricated prototype and its measurement setup in anechoic chambers.

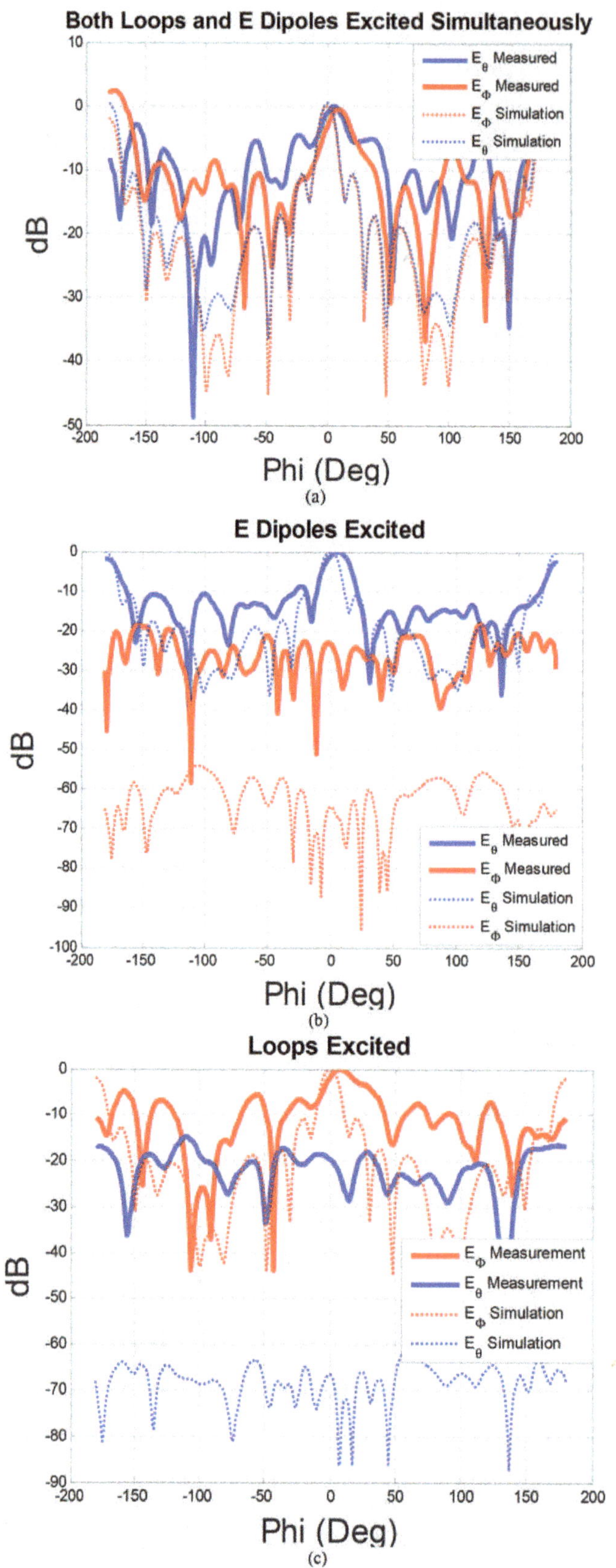

Figure 14. Illustration of co-polar and cross polar fields (amplitudes in dB) for simulation and measurement when (a) both E&M elements are excited in the array, (b) only magnetic loops are excited, and (c) only E dipoles are excited.

If each of these eight collocated units (16 elements) are excited with equal phase and equal amplitude, then it can be seen in **Figure 14(a)** that copolar H and V fields match exactly to each other which meets the requirement for weather applications (proves to be effective in extracting more hydrological parameters) and gives better resolution of the target [2]. On the other hand, if only magnetic loop is transmitting and E dipole is receiving or vice versa, the cross polar radiation is observed to be more than 70 dB lower than the peak co-polar field. These results are illustrated in **Figure 14**.

The simulation results shown above in **Figures 13** and **14** are based on the ideal environment which might differ in the case of practical construction of an array structure. The practical structure of array has feed lines, large backplane (aluminum), support structure for electronics, and other practical factors which might affect the array performance. This can actually have an adverse effect on the cross polarization levels of the array, and it is because of this we see that the simulation and measurement results of the array differ. Due to this reason, we see that the measurement results shown in **Figure 14** differ slightly from simulation. The measurement process involves extensive measures taken to achieve the best possible results, but the interferences from the feed lines and mounting structures are evident. The measurement results shown in **Figure 14** are for uncalibrated array without using any type of phase shifters for beam forming. In addition, the limitations such as availability of ideal probes which have purer linear polarization than AUT make the suppression of cross polarization more difficult.

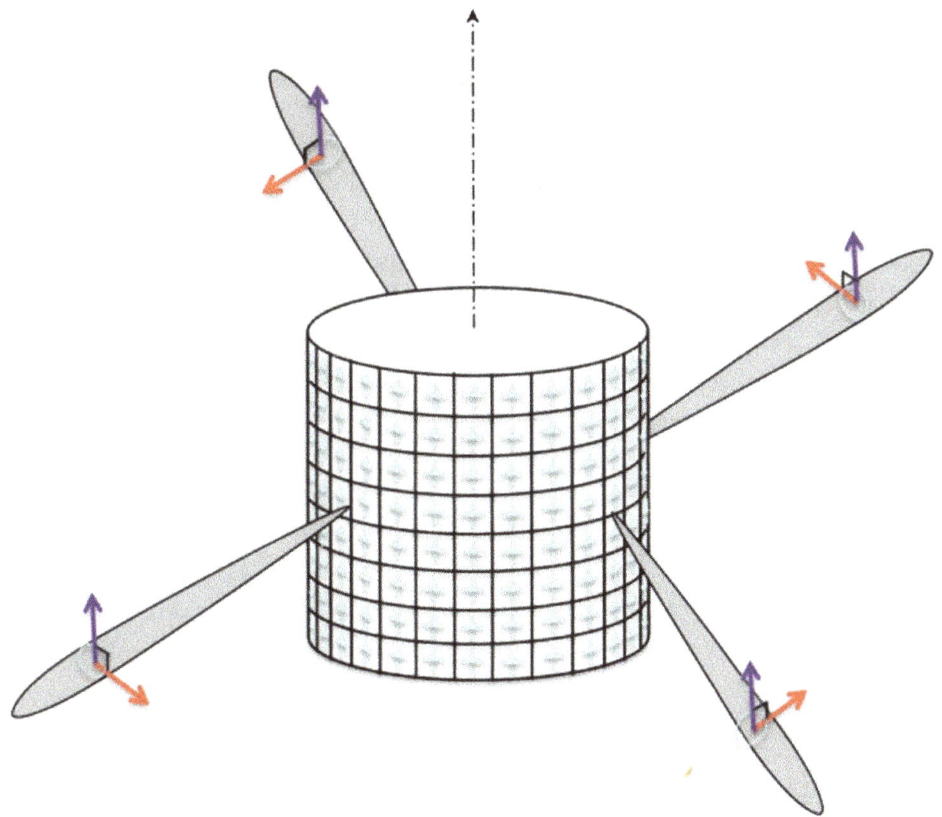

Figure 15. CPPAR with a pair of dipoles for each array element [8].

The concept of E&M dipole cannot only be implemented in linear arrays, but also extended for a cylindrical polarimetric phased array radar (CPPAR) as demonstrated in **Figure 15**. The use of CPPAR has its own advantages like scan invariant polarization basis and low sensitivity loss [8]. In addition to this, the use of the dipole elements can avoid surface waves and creeping waves. However, this is at the cost of more complex and expensive fabrication process. As aforementioned, this study is a step toward designing a new type of dual-polarized array elements based on EM dipole concept on which future research can be conducted. Initial array fabrication and testing are to be performed in a larger array test-bed called Configurable Phased Array Demonstrator (CPAD) at the Radar Innovations Laboratory (RIL), University of Oklahoma.

7. Conclusion

A dual-polarized radiating element for MPAR is proposed by using two simple types of microstrip radiating elements. As aforementioned, this work is a step toward designing and realizing a dual-polarized array element based on electromagnetic dipole concept. The intention of this chapter is to demonstrate a possible new design and build a foundation on which future research can be conducted.

This chapter shows the investigation for two different types of elements which can be collocated to form a dual-polarized element for phased array radars. The array element is a fundamental building block for an array antenna, and it is required that these elements have high cross polarization suppression capacity. The electric dipole takes the form of printed dipole. The magnetic dipole is realized by a simple loop with capacitive loading which maintains uniform surface current. An ideal electric dipole should produce only vertically polarized waves (i.e., E_θ), whereas an ideal magnetic dipole should generate only horizontally polarized waves (i.e., E_ϕ) which is free of any cross polar radiation. Nevertheless, these elements when realized practically have secondary cross polar fields due to practical factors in design and assembly and radiation from feed lines. For weather polarimetry it is expected that the elements maintain cross polarization factor of about 40 dB for simultaneous transmit/receive and 20 dB for alternate transmit/receive [6]. However, these numbers are for the array level design, but we expect that if the cross polarization isolation is minimized at the element level, then it is expected that with array level implementation, one can achieve this goal.

The measurement process involves extensive measures taken to achieve the best possible results, but the interferences from the feed lines and mounting structures are evident. The measurements performed in the anechoic chambers show the cross polarization of these elements to be about 20 dB lower than peak co-polar field. In addition, the limitations such as availability of ideal probes which have purer linear polarization than AUT make the suppression of cross polarization more difficult.

There are few aspects in which the current design of the elements can be improved. One significant improvement can be a design of a proper feed line using techniques such as proximity feed, aperture coupled feed, etc. This feeding design can be carried forward and improved by also considering the mechanical design and ease of feeding these elements when assembled into an array. It would be lucrative to see a linear array constructed using these elements such that the cross polar fields are further reduced.

Acknowledgements

Parts of this chapter are reproduced from the MS Degree Dissertation of the first author. This work is supported by NOAA-NSSL through grant #NA11OAR4320072. Any opinions, findings, conclusions, or recommendations expressed in this publication are those of the authors and do not necessarily reflect the views of the National Oceanic and Atmospheric Administration.

Author details

Ridhwan Khalid Mirza[1], Yan (Rockee) Zhang[1*], Dusan Zrnic[2] and Richard Doviak[2]

*Address all correspondence to: rockee@ou.edu

1 Intelligent Aerospace Radar Team, Advanced Radar Research Center, School of ECE, University of Oklahoma, Norman, USA

2 National Severe Storm Laboratory, NOAA, Oklahoma, Norman, USA

References

[1] NOAA National Severe Storms Laboratory [Online]. Available: http://www.nssl.noaa.gov/tools/radar/mpar/, 2016, 20th January 2016.

[2] D. S. Zrnic, V. M. Melnidov, and R. J. Doviak, "*Issues and challenges for polarimetric measurement of weather with an agile beam phased array radar*," NOAA/NSSL report, [Updated: 2 May 2013], 2013.

[3] C. E. Baum, "Symmetry in electromagnetic scattering as a target discriminant," in *Optical Science, Engineering and Instrumentation'97*, 1997, pp. 295-307.

[4] D. Vollbracht, "Understanding and optimizing microstrip patch antenna cross polarization radiation on element level for demanding phased array antennas in weather radar applications," *Advances in Radio Science*, vol. 13, pp. 251-268, 2015.

[5] Kuloglu, Mustafa. "Development of a Novel Wideband Horn Antenna Polarizer and Fully Polarimetric Radar Cross Section Measurement Reference Target." PhD diss., The Ohio State University, 2012. [Web]. Available:https://etd.ohiolink.edu/rws_etd/document/get/osu1338387100/inline

[6] Y. Wang and V. Chandrasekar, "Polarization isolation requirements for linear dual-polarization weather radar in simultaneous transmission mode of operation," *Geoscience and Remote Sensing, IEEE Transactions on*, vol. 44, pp. 2019-2028, 2006.

[7] F. Mastrangeli, G. Valerio, A. Galli, A. De Luca, and M. Teglia, "An attractive S-band dual-pol printed antenna for multifunction phased array radars," in *Antennas and Propagation (EUCAP), Proceedings of the 5th European Conference on*, pp. 514-516, 2011.

[8] G. Zhang, R. J. Doviak, D. S. Zrnic, R. Palmer, L. Lei, and Y. Al-Rashid, "Polarimetric phased-array radar for weather measurement: a planar or cylindrical configuration?," *Journal of Atmospheric and Oceanic Technology*, vol. 28, pp. 63-73, 2011.

[9] C. A. Balanis, *Antenna theory: analysis and design* vol. 1: John Wiley & Sons, 2005.

[10] W. J. Wu, R. Fan, Z. Y. Zhang, W. Zhang, and Q. Zhang, "A shorted dual-polarized cross bowtie dipole antenna for mobile communication Systems," in *General Assembly and Scientific Symposium (URSI GASS), 2014 XXXIth URSI*, pp. 1-4, 2014.

[11] K. M. Mak, H. Wong, and K. M. Luk, "A shorted bowtie patch antenna with a cross dipole for dual polarization," *IEEE Antennas and Wireless Propagation Letters*, vol. 6, pp. 126-129, 2007.

[12] Y. Liu, H. Yi, F. W. Wang, and S. X. Gong, "A novel miniaturized broadband dual-polarized dipole antenna for base station," *IEEE Antennas and Wireless Propagation Letters*, vol. 12, pp. 1335-1338, 2013.

[13] A. Ludwig, "The definition of cross polarization," *Antennas and Propagation, IEEE Transactions on*, vol. 21, pp. 116-119, 1973.

[14] D. Acosta. *Enriched Physics 2 Lecture* [Web]. Available: http://www.phys.ufl.edu/~acosta/phy2061/lectures/MagneticDipoles.pdf, 2016, 23rd February

[15] J. D. Kraus, Antennas, Tata McGraw Hill Edition,1988. [Web]. Available:http://117.55.241.6/library/E-Books/Antennas_mcgraw-hill_2nd_ed_1988-john_d_kraus.pdf.

[16] D. Foster, "Loop antennas with uniform current," *Proceedings of the IRE*, vol. 32, pp. 603-607, 1944.

[17] R. Li, N. Bushyager, J. Laskar, and M. Tentzeris, "Circular loop antennas reactively loaded for a uniform traveling-wave current distribution," in *Antennas and Propagation Society International Symposium, 2005 IEEE*, pp. 455-458, 2005.

[18] K. Wei, Z. Zhang, and Z. Feng, "Design of a wideband horizontally polarized omnidirectional printed loop antenna," *Antennas and Wireless Propagation Letters, IEEE*, vol. 11, pp. 49-52, 2012.

[19] J. H. Park, Y. H. Ryu, and J. H. Lee, "Mu-zero resonance antenna," *Antennas and Propagation, IEEE Transactions on*, vol. 58, pp. 1865-1875, 2010.

[20] *Dipole Antenna* [Web]. Available: https://en.wikipedia.org/wiki/Dipole_antenna, 2016, 5th March 2016.

[21] D. M. Pozar, *Microwave engineering*: John Wiley & Sons, 2009.

[22] D. J. P. Donohoe. (23 rd February), Mississippi State University. [Web].Available:http://my.ece.msstate.edu/faculty/donohoe/ece4990notes5.pdf

5

Application of Composite Right/Left-Handed Metamaterials in Leaky Wave Antennas

Keyhan Hosseini and Zahra Atlasbaf

Abstract

This chapter reviews the most significant advancements in the context of metamaterial (MTM) leaky wave antennas (LWAs). A brief review of the mechanism of leaky wave radiation along with an important class of MTMs known as composite right/left-handed (CRLH) structures is presented. Then, recent outstanding works in the area of CRLH LWAs are reported in detail. These works include the application of electronic control, substrate integrated waveguides, dual band and wideband performance, ferrite loaded waveguides, and split-ring-resonator (SRR)-based MTMs in LWAs. Also, the benefits of LWAs to design high gain active structures, reflecto-directive systems, wideband dual-layer substrate integrated waveguide antennas and conformal antennas are discussed.

Keywords: composite right-left handed (CRLH), dispersion diagram, leaky wave antenna (LWA), metamaterial (MTM), transmission line (TL)

1. Introduction

A leaky wave is a traveling wave progressively leaking out power as it propagates along a waveguiding structure. A leaky wave antenna (LWA) supports leaky waves where the leakage phenomenon is generally associated with high directivity [1]. LWAs are different from resonating antennas, as they are based on traveling waves as opposed to resonate-wave mechanism. Hence, their size is not related to the operation frequency, but to directivity [2]. A schematic LWA is depicted in **Figure 1**. If the wave is faster than the velocity of light ($\beta < k_0$), k_y is real and the leakage radiation occurs. This wave is called a fast wave. In this case, β determines the angle θ_{mb} of radiation of the main beam and as illustrated in **Figure 1**, follows the simple angle-frequency relation [1].

$$\theta_{mb} = \sin^{-1}\left(\beta / k_0\right) \tag{1}$$

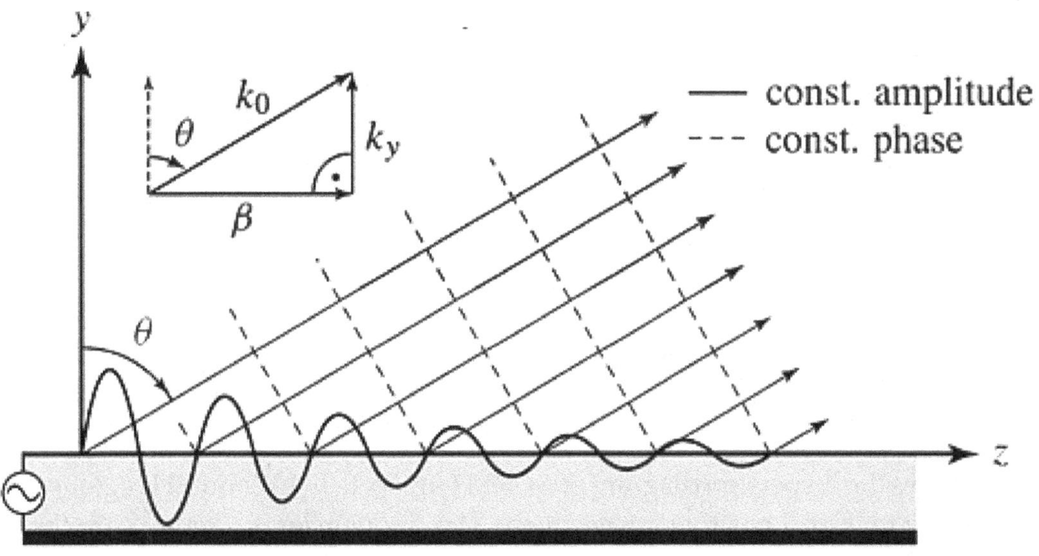

Figure 1. A general schematic LWA. The solid and dashed lines represent the constant amplitude and constant phase planes, respectively [3].

The values $\beta < 0, \beta = 0$, and $\beta > 0$ correspond to backward, broadside and forward angles, respectively. If the waveguide is dispersive (β is a nonlinear function of ω) the main beam angle is changed as a function of frequency, $\theta_{mb} = \theta_{mb}(\omega)$. This phenomenon is called *frequency scanning*. Eq. (1) also reveals that radiation in any angle from backfire ($\theta_{mb} = -90$ °) to endfire ($\theta_{mb} = +90$ °) can be potentially achieved if β varies from $-k_0$ to $+k_0$.

Right-handed (RH) structures have limitations in space scanning. In non-periodic RH LWAs, $\beta > 0$ at all frequencies and hence, only forward angles can be obtained. In periodic RH LWAs, there are infinite number of space harmonics β_n, some of which are fast waves. The harmonics can support both backward ($\beta_n < 0$) and forward ($\beta_n > 0$) waves. However, the systematic presence of a gap at $\beta = 0$ prevents broadside radiation. A left-handed (LH) or MTM leaky wave antenna does not face the above-mentioned constraints and supports a continuous backward to forward frequency scanning. Also, since its fundamental mode is a leaky wave, no lower order guiding mode is to be suppressed and hence no complex feeding structure is needed [4].

Purely LH structures do not exist in nature since there are also RH parasitic effects. Such structures are known as composite right/left-handed (CRLH) media, whose typical transmission line (TL) model is depicted in **Figure 2(a)**. After simple manipulations, the propagation constant γ of a CRLH structure can be obtained as

$$\gamma = \alpha + j\beta = \pm j\sqrt{\left(\omega L_R - \frac{1}{\omega C_L}\right)\left(\omega C_R - \frac{1}{\omega L_L}\right)}. \tag{2}$$

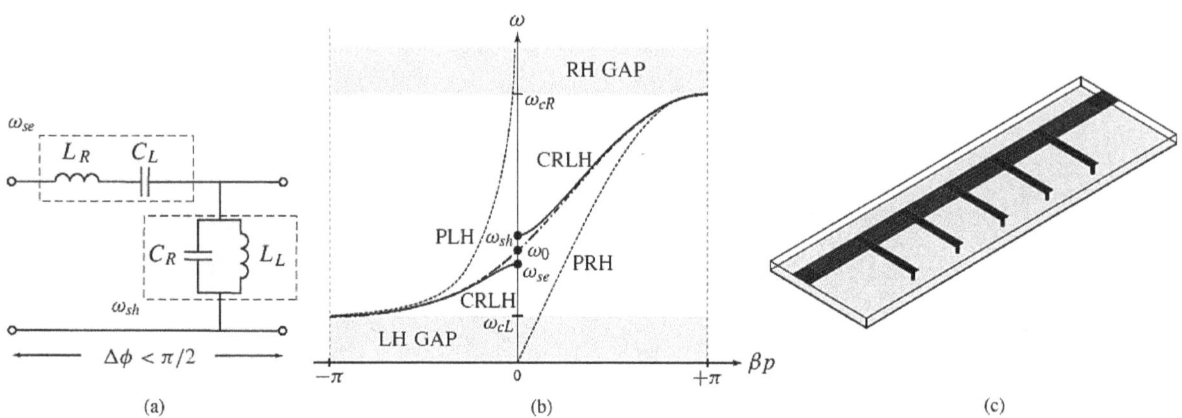

Figure 2. CRLH unit cell, (a) a typical TL model. The pairs (L_R, C_R) and (L_L, C_L) represent the RH and LH natures of the medium, respectively, (b) dispersion diagram for unbalanced case, and (c) interdigital capacitor-shunt stub implementation of CRLH unit cell [4].

Figure 2(b) shows the dispersion diagram of a CRLH unit cell. In LH and RH regions, the phase constant β is negative and positive, respectively. The frequencies ω_{se} and ω_{sh} are the through and shunt branches resonant frequencies, respectively. In a *balanced* CRLH TL, $\omega_{se} = \omega_{sh}$. In this case, the *transition* frequency ω_0 is equal to either of the resonant frequencies and there exists a broadside radiation if there is a leaky wave. In an unbalanced CRLH TL $\omega_{se} \neq \omega_{sh}$, and there exists a stop-band in this frequency range. In the dispersion diagram, the leakage radiation region ($\beta < k_0$), corresponds to inner region of a cone (radiation cone). To design a LWA the operation frequency is chosen to lie in the radiation cone. The frequencies ω_{cL} and ω_{cR} are cut-off frequencies for LH and RH regions, respectively. Note that in the stop-bands, $\beta = 0$ and from Eq. (2), $\gamma = \alpha$.

Because of their non-resonant nature, CRLH structures can be designed to exhibit simultaneously low loss and broad bandwidth. Low loss is achieved by a balanced design ($L_R C_L = L_L C_R$) of the structure and good matching to the excitation ports, whereas broad bandwidth is a direct consequence of the TL nature of the structure. Another advantage of CRLH MTMs is that they can be engineered in planar configurations, compatible with modern microwave integrated circuits [5]. A practical planar implementation of the CRLH unit cell is shown in **Figure 2(c)**. It consists of an interdigital capacitor as a series LH capacitor and a shorted shunt stub as a shunt LH inductor. Obviously, there are always parasitic RH capacitors and inductors due to the coupling between metallic trace and the ground plane. This forms a CRLH unit cell.

In the following, the most significant recent works in the context of CRLH LWAs are reviewed.

2. Electronically controlled CRLH LWA with tunable radiation angle and beamwidth [6]

Radiation angle electronic control can be explained in terms of Eq. (2). From the equivalent circuit of **Figure 2**, the dispersion diagram is obtained. Hence, from Eq. (1), the capacitors and inductors determine the scan angle. If the one can control the capacitors by the voltage, the angle of radiation is controlled by the bias voltage, accordingly. This fact is shown in **Figure 3**.

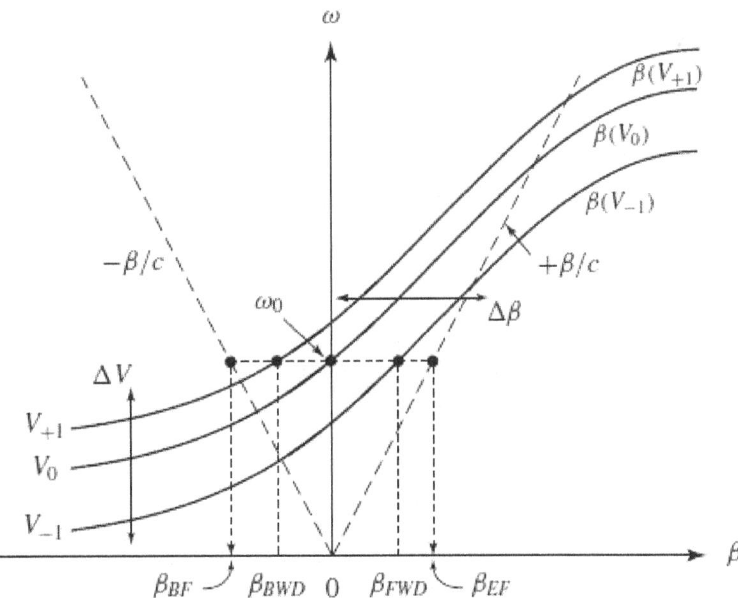

Figure 3. Principle of beam electronic scanning. Dispersion curves are shifted vertically as bias voltages are varied [6].

At a fixed frequency, the varactor diodes may control the antenna operation by a proper bias voltage tuning. The capacitance of such diodes changes by tuning the reverse bias voltage and hence, β behaves as a function of reverse bias. From **Figure 3**, by tuning the bias voltage at a fixed frequency a vertical shift happens in the dispersion diagram.

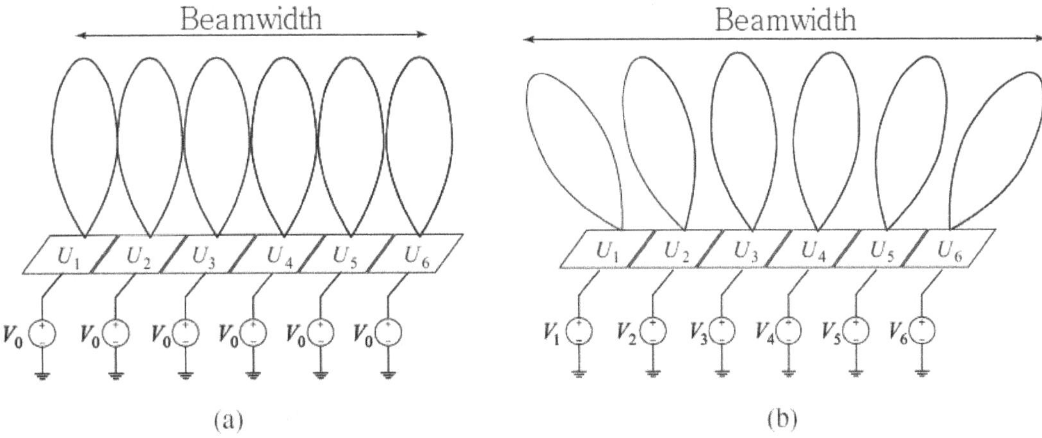

Figure 4. Principle of beamwidth electronic tuning at nonuniform biasing condition, (a) Narrow beamwidth by introduction of uniform biasing ($V_1 = V_2 = \ldots = V_0$), and (b) wide beamwidth by introduction of nonuniform biasing ($V_1 \neq V_2 \neq \ldots \neq V_0$) [6].

A uniform bias distribution along the transmission line results in a constant β and hence the LWA radiation pattern has a maximum directivity. In practice, the modification of β is difficult.

Via non-uniform biasing, the β factor of unit cells can be tuned using varactor diodes. The concept is described in **Figure 4**. When the bias distribution is uniform (**Figure 4a**), each unit cell radiates toward the same angle and the structure functions as a common LWA controlling the radiation angle. In the case of a non-uniform periodic TL, i.e., **Figure 4b**, there are different radiation angles corresponding to each cell and the beamwidth of the LWA can be controlled.

Figure 5(a) shows the layout and equivalent circuit of the voltage-controlled CRLH TL unit cell. The CRLH transmission line is loaded with varactors. One shunt and two series varactors contribute to a unit cell. A simple bias network is used to feed unit cells which, practically, can be implemented much easier than its similar counterparts. Assuming a lossless case, varactor diodes are simple modeled as a series inductor and series capacitor. **Figure 5(b)** shows the fabricated 30-cell periodic TL structure which is built on RT/Duroid5880 (ε_r = 2.2, h = 62 mil). Metelics MSV 34060-E28X Si abrupt varactor diodes are periodically distributed, and Murata chip inductors with 4.7 nH are used for dc feeds. The total length of the structure is 38.34 cm operating as a 3.3-GHz LWA. Port 1 is used for the input and Port 2 is terminated with 50 Ω in order to suppress undesired spurious beams due to mismatch reflection.

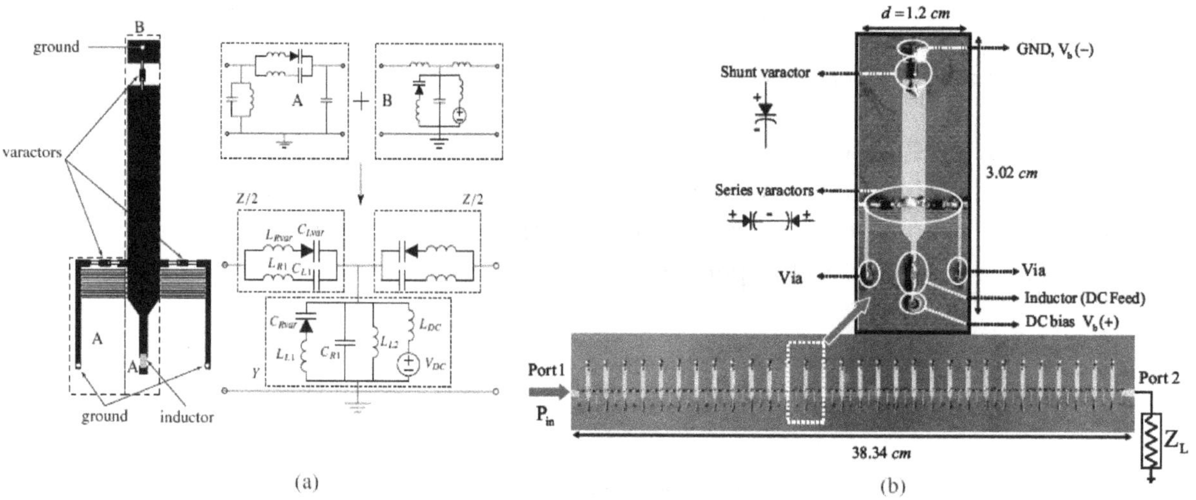

Figure 5. (a) Layout and circuit model for the varactor-loaded CRLH unit cell and (b) fabricated varactor-loaded LWA [6].

3. CRLH substrate integrated waveguide leaky-wave structure [7]

Half mode substrate integrated waveguide (HMSIW) antennas developed in [7] can support backfire to endfire radiation angle and also, its implementation is practically easy. These kinds of antennas may be fabricated by creating interdigital slots on the metal trace and ground plane of the structure. The vias and the slots perform as shunt inductor, and series capacitor, respectively. This, in turn, leads to a radiation in backward direction. The LWAs can be miniaturized by forcing them operate in the frequencies below the cutoff frequency. Also, half of the structure is utilized which makes the antenna smaller in the operating frequency.

Figure 6(a) shows the configurations of the one period CRLH SIW element. Single side and double side radiations are possible. **Figure 6(b)** shows the unit cells of CRLH HMSIW structure. For the conventional HMSIW, because of the large width-to-height ratio and the metallic via array, only the quasi-TE$_{p-0.5,0}$ ($p = 1, 2, \ldots$) modes are able to propagate in the guiding structure. In this antenna, the guided- and radiated-wave operation is fulfilled at the frequencies above and below the cutoff frequency of the waveguide. In the right unit cell demonstrated in **Figure 6(b)**, a wall constituted by metallic vias with a perfect electric conductor (PEC) strip on the wall is located. This wall is utilized to decrease the power that may leak from the boundary of the LWA. The above mentioned transmission lines can be easily mounted on the metallic surface.

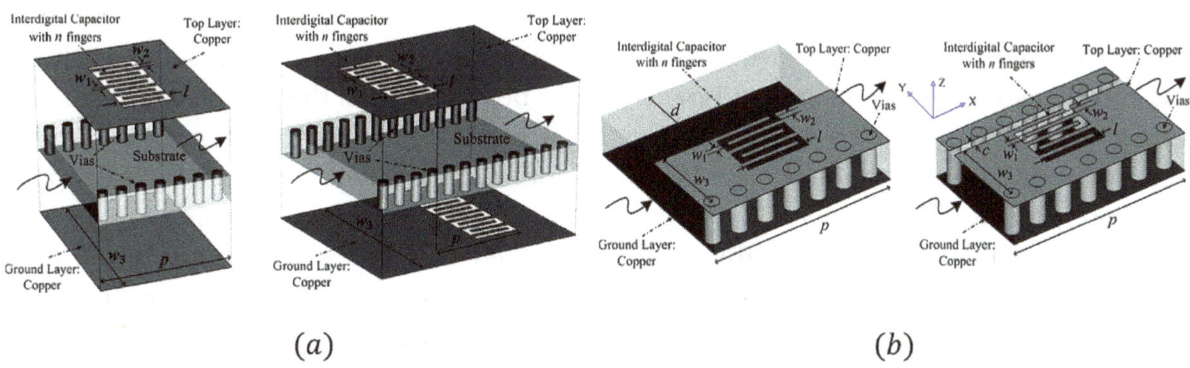

Figure 6. (a) Single side and double side radiating elements in the SIW LWA and (b) the initial unit cell and the modified folded unit cell for the CRLH HMSIW LWA [7].

Two X-band LWAs fabricated in [7] are shown in **Figure 7**. The substrate is Rogers 5880 with a thickness of 0.508 mm and relative permittivity of 2.2. **Figure 7(a)** shows one-sided and double-sided SIW LWAs. **Figure 7(b)** shows two HMSIW LWAs with unbalanced CRLH unit cells. The lower antenna is the modified folded HMSIW LWA.

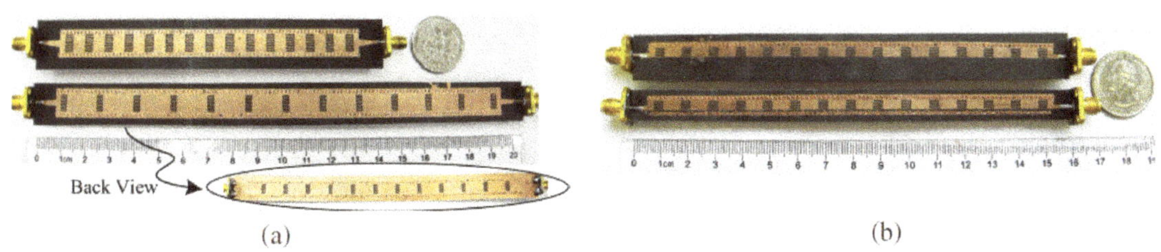

Figure 7. Fabricated LWAs, (a) SIW configuration and (b) HMSIW configuration [7].

Figure 8(a) shows the simulated and measured transmission responses of the folded HMSIW LWA. It is unbalanced and a bandgap is observed, since near the transition frequency of the CRLH unit cell the return loss is larger than its acceptable threshold.

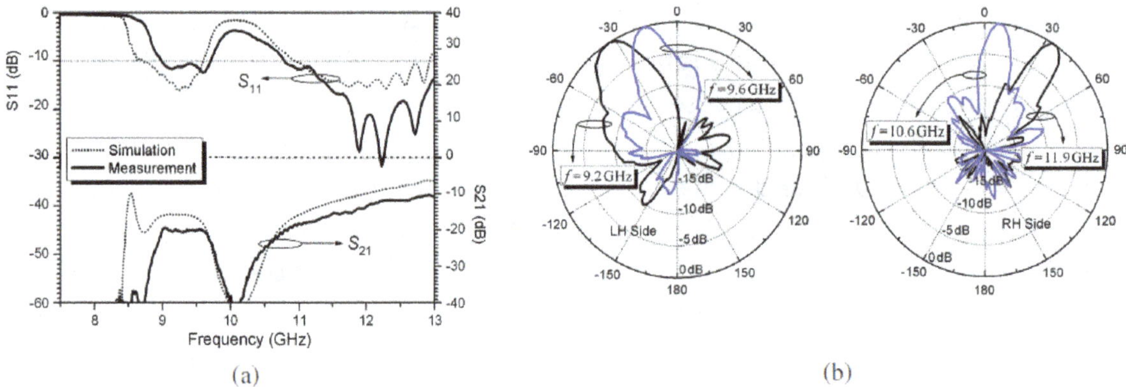

Figure 8. (a) Measured and simulated S-parameters and (b) measured E-plane radiation patterns in LH- and RH-regions, for the folded HMSIW LWA [7].

The discrepancy between the measurement and simulation is due to the fact that the conductor loss and dielectric loss in the measurement should be higher than that in the simulation. By changing the slot size and the position of the vias the position of the LH band can be controlled. Balanced condition can also be obtained by some optimizations. **Figure 8(b)** shows the E-plane radiation patterns for the folded CRLH HMSIW LWA. Beam frequency-scanning capability in E-plane is clearly observed. Since this is an unbalanced case and there is no balanced point which gives broadside radiation it is difficult to obtain the H-plane patterns. The edge radiation in HMSIW LWA is slightly higher compared with the SIW LWA, which is mainly due to edge radiation caused by the open boundary.

4. High gain active CRLH LWA [8]

In the reference [8], a high-gain CRLH LWA is presented to operate at broadside. It is constituted by passive CRLH leaky-wave sections interconnected by amplifiers, which regenerate

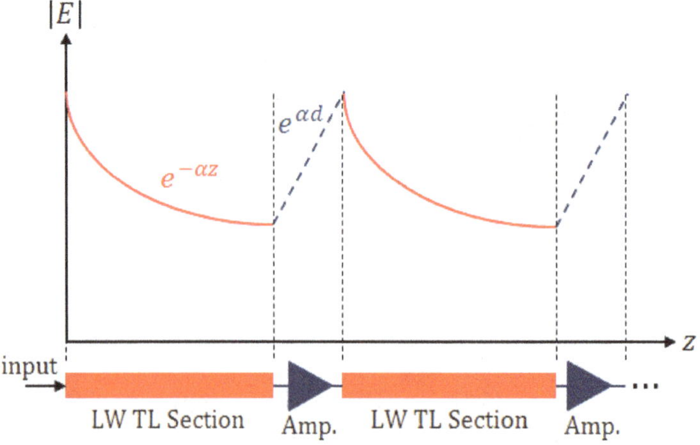

Figure 9. In an active LWA, the field intensity (exponential) loss is compensated by amplification [8].

the power progressively leaked out of the structure to increase the effective aperture of the antenna and thereby its gain. The gain is further enhanced by a matching regeneration effect induced by the quasi-unilateral nature of the amplifiers. The principle of gain enhancement is shown in **Figure 9**.

Practically, a matched load is located at the end of the leaky-wave antennas. The power has leaked by 90% before reaching the termination point since no more increase in the radiation area of the antenna is no more achieved. On the other hand, if the guided wave is boosted periodically, the gain of the LWA can be arbitrarily increased. Another advantage is that this provides with the new possibility of achieving arbitrary current distributions along the antenna metallization and henceforth generating arbitrary radiation patterns. The fabricated active LWA is shown in **Figure 10(a)**. As shown in **Figure 10(b)**, the amplifier is on the bottom layer and is connected to the top microstrip line through vertical transition of vias. The amplification loop is designed to provide a zero phase shift from its input to its output in order to play a neutral role in terms of phase at the transition frequency ($\beta = 0$).

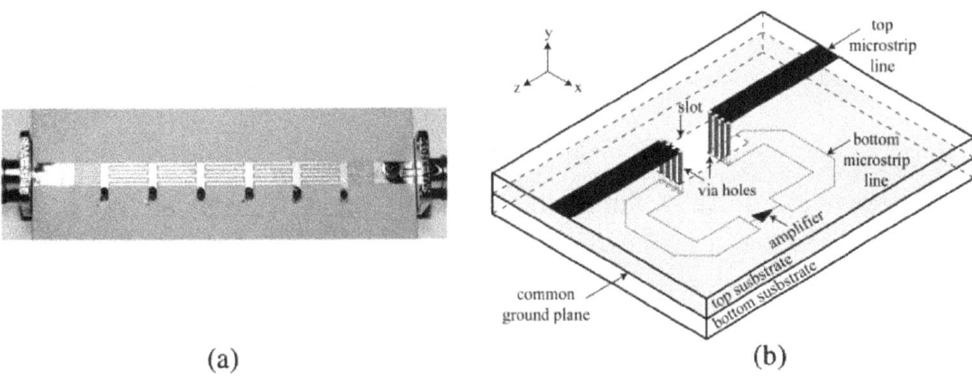

Figure 10. (a) The fabricated active LWA and (b) microstrip-to-microstrip vertical double transition to embed the amplifier in the unit cell [8].

5. Reflecto-directive system using CRLH LWA and heterodyne mixing [9]

In the reference [9] a system receives a signal at a fixed frequency with a patch antenna and reflects it back toward any desired angle by tuning the LO frequency of a mixer. The principle of operation is based on the reception of a wave at a fixed frequency by a quasi-omnidirectional microstrip receiver and retransmission of it to an arbitrary direction by a full space scanning LWA. One can accomplish the new function by way of heterodyne mixing. A schematic of the reflecto-directive system is shown in **Figure 11(a)**. In this schematic, two mixers are required if f_{in} lies in the band of the LWA. In the case of utilizing only one mixer, the incoming wave cannot be filtered and hence it contributes to generation of an undesired angle existing at the input frequency. The mentioned fact is practically cause trouble to the reflecto-directive system and is maintained using a circuit including two mixers, where the signal at the input frequency is removed using a low-pass filter placed just before the second mixer. The reflecto-directive system prototype is shown in **Figure 11(b)**.

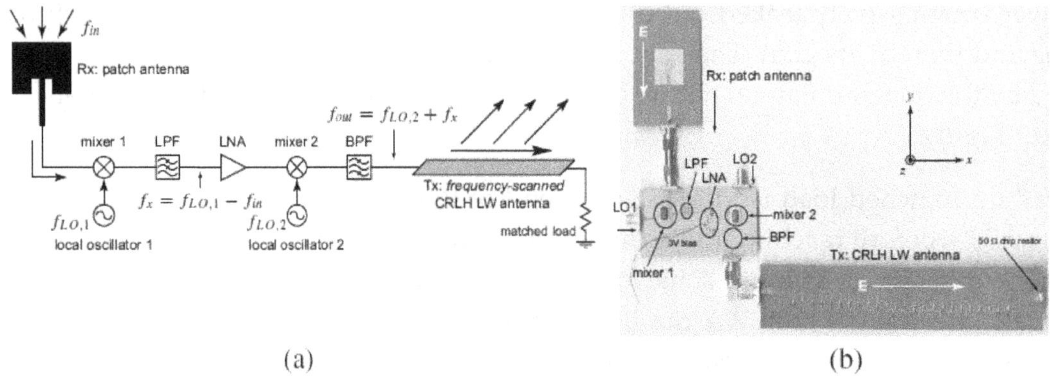

Figure 11. The reflecto-directive system, (a) schematic and (b) prototype. The polarization of RX and TX antennas are along *y*-direction [9].

The source test antenna sends a signal at the frequency of 5 GHz at two different fixed angles θ_{in} = − 30 °, 30 ° .The radar cross section of the reflecto-directive system is measured in the frequencies, f_{out} = 3.5 GHz, 4.5 GHz and 5.5 GHz. Based on the following relation, the corresponding $f_{LO,2}$ for each input frequency can be obtained.

$$f_{LO,2} = f_{out} + f_{in} - f_{LO,1}. \quad (3)$$

The results are shown in **Figure 12.** For all the incidence angles tested, clear radar cross section (RCS) peaks appear at the angles −20°, +20°, and +50° for the frequencies f_{out} = 3.5 GHz, 4.5 GHz and 5.5 GHz, respectively, in agreement with the LWA angle-frequency relation.

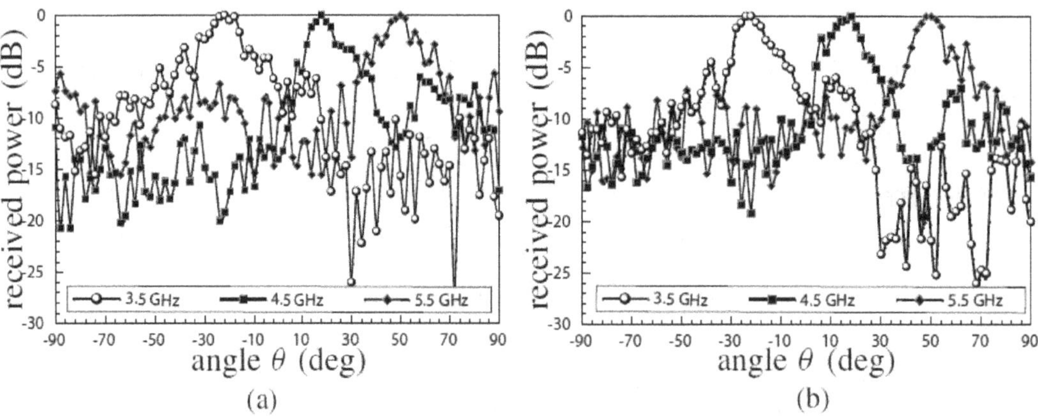

Figure 12. Measured bistatic RCS of the reflecto-directive system [9].

6. Dual band LWA on a CRLH substrate integrated waveguide [10]

The idea of the antenna design in [10] is based on dual band/quad band operation of the CRLH TL. The layout of the proposed dual band SIW LWA in [10] is shown in **Figure 13**. In its unit cell equivalent circuit, the series elements are represented mainly by the meander slots that enable the antenna to radiate, and the parallel elements are represented by four conducting vias.

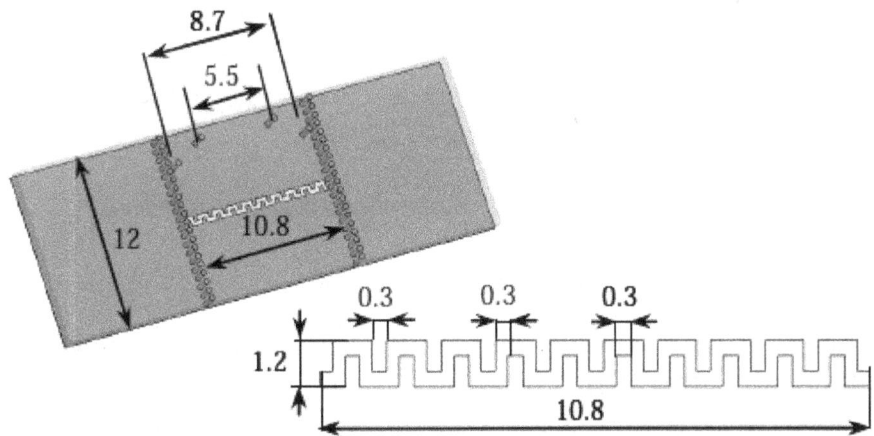

Figure 13. Layout of the meander slot and the SIW unit cell setup [10].

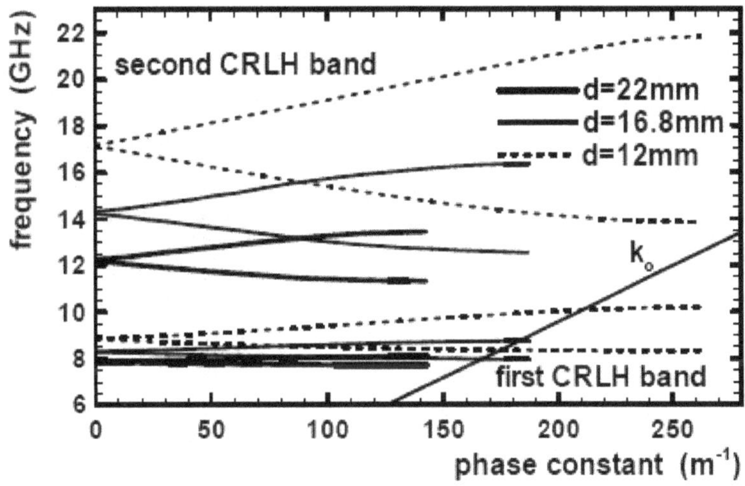

Figure 14. Dispersion diagram of the dual-band LWA for different cell lengths. The three antennas also differ in SIW width [10].

An undesired radiation exists in the antenna since in the structure when the relation $\beta d = \pi$ is satisfied, the Bragg phenomenon takes place. Bragg reflection occurs when the dispersion diagram is at its edges. The dispersion diagrams of the antennas are demonstrated in **Figure 14**. The unit cell lengths in these antennas are different. **Figure 14** shows that the central frequency of the higher frequency band is controlled mainly by the length of the unit cell and the central frequency of the lower band is mainly determined by the SIW width. If $\beta d = \pi$, spurious radiation occurs. This can be removed by shifting this point into a non-radiating area, or at least to the close proximity of it. The drawback is of course that the steering beam span is reduced. From **Figure 14**, the antenna dual band operation is clearly observed. A photograph of the 25-cell antenna is shown in **Figure 15**.A quarter-wavelength transformer was used to transform the real part of Z_{in} to nearly the same value as the impedance of the empty SIW. The transformer is located at the point where the imaginary part of Z_{in} is almost equal to zero. The main advantage of this type of matching is that it matches the first band of the antenna, but it simultaneously does not affect the second band.

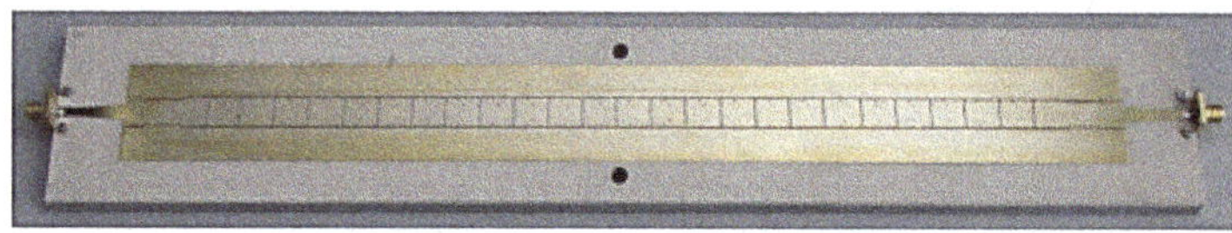

Figure 15. The fabricated dual-band SIW LWA [10].

The measured radiation patterns of the antenna in the longitudinal plane normal to the antenna are shown in **Figure 16**. Part (a) and (b) of **Figure 16** correspond to lower- and higher-frequency bands of the antenna, respectively. Steering the direction of the main lobe from backward to forward by changing frequency in the two frequency bands is obvious. In the second band, the pencil beam is narrower but the beam steering is less sensitive as the beam spans a smaller angular range.

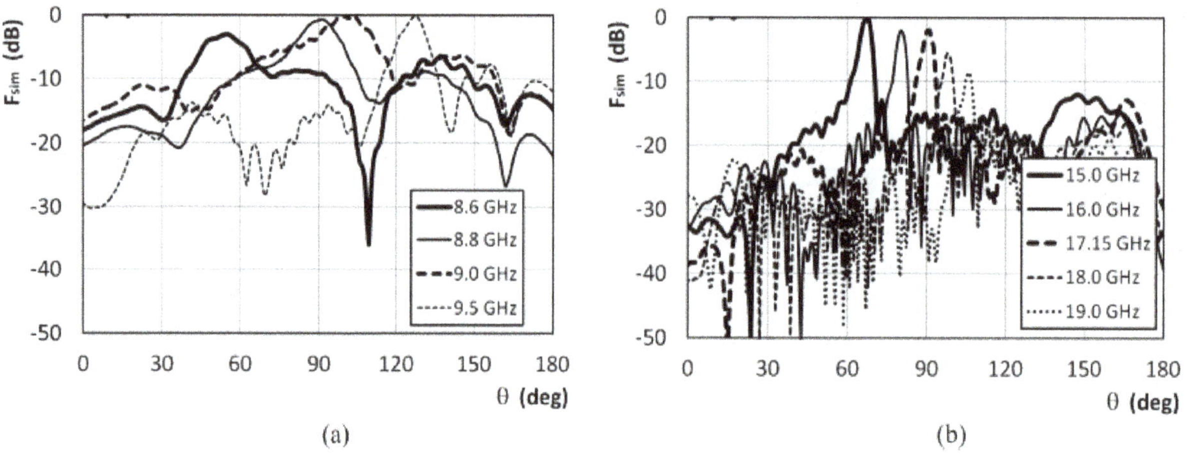

Figure 16. Measured radiation patterns, (a) first band and (b) second band [10].

7. Dual band full space scanning LWA based on ferrite loaded waveguide [11]

The ferrite-loaded open structure is shown in **Figure 17(a)**. It is a waveguide fully filled with ferrite with one side wall removed and magnetized by the bias field H_0. At the frequency band in which n_e is much larger than one ($n_e \gg 1$), the side wall of the structure which is the interface between vacuum and magnetized ferrite, can be assumed as a perfect magnetic conductor (PMC) medium.

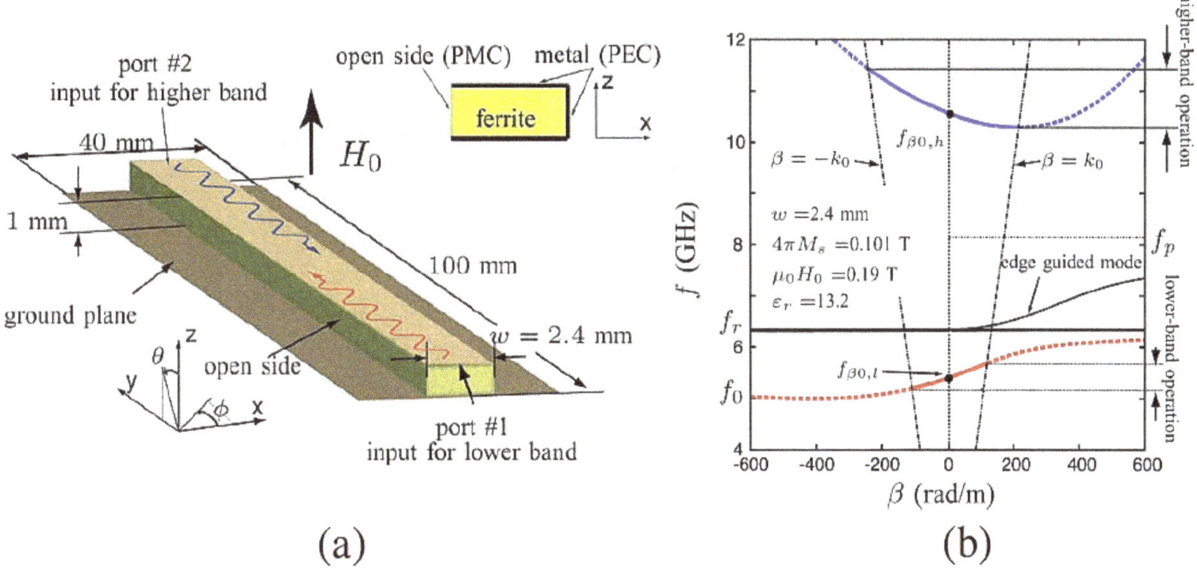

Figure 17. (a) Ferrite-loaded open waveguide structure and (b) its dispersion diagram [11].

The dispersion relation of the structure in **Figure 17(b)** is

$$\tan(k_x w) = -\frac{\mu k_x}{\beta \kappa}, \tag{4}$$

with $k_x = \sqrt{\omega^2 \varepsilon \mu_e - \beta^2}$, and μ and κ are the elements of Polder tensor permeability, and μ_e is the effective birefringent permeability of the structure, corresponding to a bias field H_0 perpendicular to the direction of propagation β. The dispersion diagram computed by Eq. (4) is shown in **Figure 17(b)**. Essentially, three bands may be observed in the frequency range shown. The lower band is the CRLH mode, the second band is the edge-mode band, and the higher band is called mixed forward/backward perturbed TE_{10} mode. In this LWA only the leaky modes are of interest. These modes are present at the lowest and the highest bands.

From $v_g = d\omega/d\beta$, the lower band CRLH leaky mode exhibits a positive group velocity v_g at all frequencies. Since $v_g > 0$, power flows in the positive y-direction. In contrast, due to non-reciprocity of the structure and the subsequent nonexistence of a $v_g < 0$ mode, no energy can propagate toward the source in this mode, even in the case of reflection, where power would be dissipated in heat instead of propagating. The phase velocity of the lower-band CRLH mode changes sign from negative to positive at the transition frequency when frequency is increased. From the scanning law for the main leaky-wave beam, the structure exhibits full-space radiation from backfire to endfire. The higher band leaky mode exhibits a negative group velocity at the frequencies within the leaky-wave region, which is the region of interest. The phase velocity of the higher perturbed waveguide mode is negative and positive below and above the transition frequency leading to full space beam scanning.

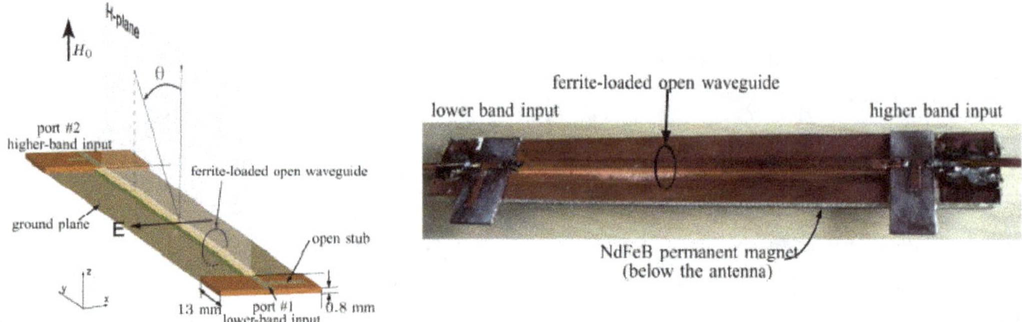

Figure 18. The configuration and prototype of the ferrite-loaded open waveguide [11].

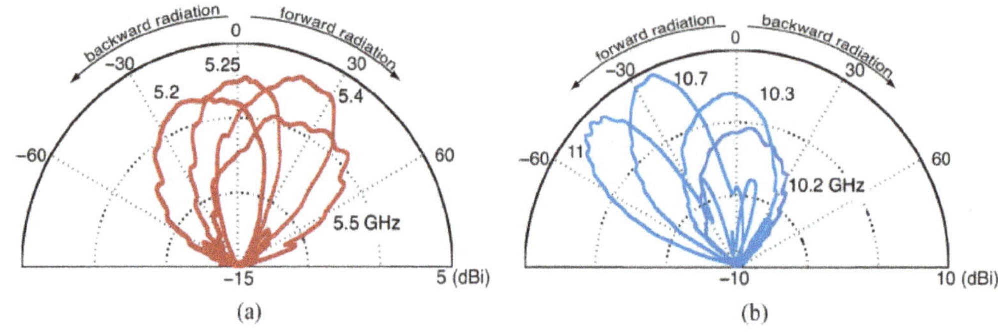

Figure 19. Measured frequency beam scanning, (a) lower band and (b) higher band [11].

Figure 18 shows the configuration and the prototype of the dual band antenna with microstrip-to-waveguide transitions and lower-band/higher-band stub matching sections at both ends of the structure. The measured radiation patterns are shown in **Figure 19**. Note that due to the fact that the ports of the two modes are placed at the opposite ends, the forward and backward spaces are reversed between the two modes.

8. Fundamental-mode LWA using slot-line and split-ring-resonator (SRR)-based MTMs [12]

The key point in [12] in the fact if the slots in a coplanar waveguide (CPW) TL have a relatively far distance from each other, the structure is regarded as two independent TLs loaded with slots both lying on a dielectric substrate, which have two magnetic currents with 180 ° phase shift. The propagation characteristics of the CRLH CPW TL based on split-ring-resonators (SRRs) and wires can be obtained through the analysis of the dispersion relation. This can be inferred from the lumped element equivalent circuit model of **Figure 20(a)**. The implementation of the CRLH slotline LWA is shown in **Figure 20(b)**. The antenna is essentially a host slot-line periodically loaded with SRRs and narrow metallic strips. The unit cell was obtained by removing one of the halves of the CPW unit cell. The loaded line is matched to a connector using a semi-lumped microstrip-to-slotline transition acting as an inductive transformer-based matching network.

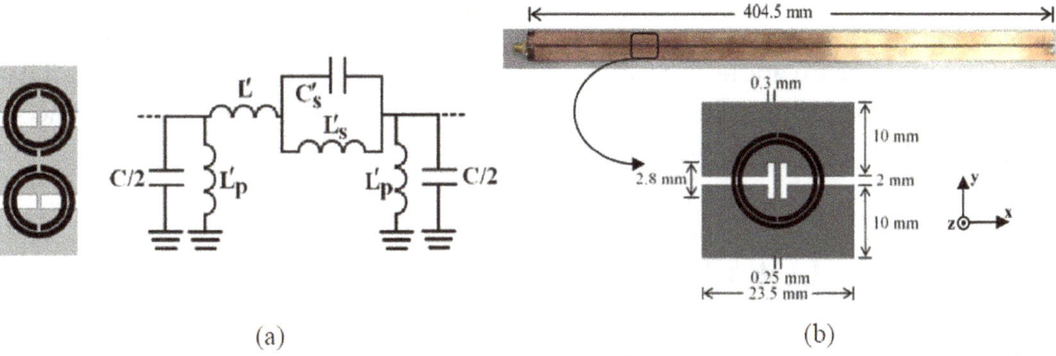

Figure 20. (a) Layout of the unit cell of the CPW structure with SRRs (etched in the back substrate side) and its simplified equivalent circuit and (b) fabricated LWA [12].

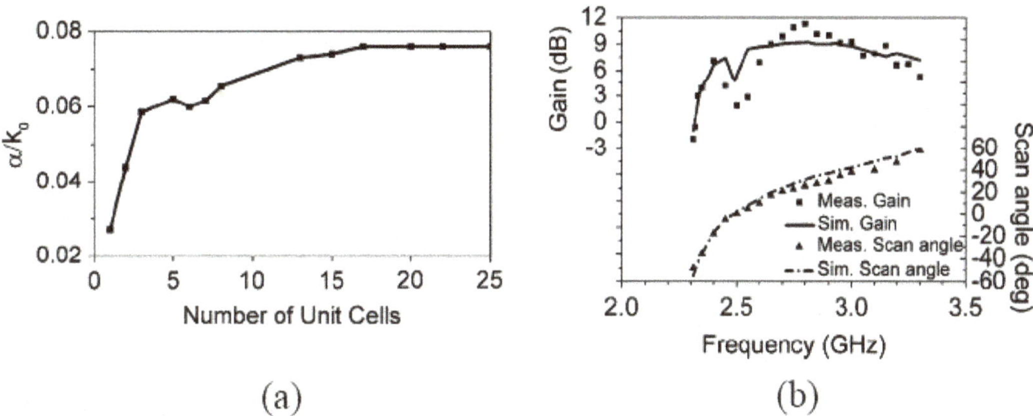

Figure 21. The SRR-based LWA, (a) attenuation constant versus the number of unit cells at 2.5 GHz and (b) gain and scan angle [12].

Because of edge effects and mutual coupling, the attenuation constant (α)of an LWA cannot be obtained through the analysis of a unit cell. In this case, if the number of cells increases, one can neglect the edge effects and calculation of α become more accurate. Demonstrated in **Figure 21(a)** is α/k_0 which is normalized to the free space wavenumber. As seen, at least 15 unit cells are needed to reach the value $\alpha/k_0 = 0.076$. The obtained attenuation constant is utilized to derive the length of leaky wave antenna using relation Eq. (5) as stated below:

$$P_n = P_0 e^{-2\alpha n d}, \quad (5)$$

where d is the periodicity, P_0 is the power delivered to the LWA and P_n is the power at the nth terminal. For 95% of the power dissipated before reaching the antenna termination and using the converged value of $\alpha/k_0 = 0.076$, the needed number of unit cells is obtained to be $N = 17$.

Figure 21(b) shows the measured and simulated LWA gains. For the LH and RH frequency bands, the maximum measured gains are respectively 7.1 and 11.3 dB. Since the LWA balance condition at the transition frequency is not satisfied, the gain reduces at broadside angle. This results in a quite wide main beam at broadside. The existence of an evanescent mode at 2.5 GHz makes shorter the effective aperture length of the LWA and hence imperfect matching at broadside. The main beam angle as a function of frequency is also plotted in **Figure 21(b)**. The backward to forward scanning ranges from – 50 ° to + 60 ° while maintaining an acceptable gain level.

9. Conformal CRLH LWA [13]

In the reference [13], a conformal mapping-based method to analyze and synthesize singly-curved microstrip structures is reported. Conformal mapping is a powerful method which has been applied to study the guided- and radiated-wave characteristics of microstrip patch antennas [14–16]. Also, it is applied to CRLH LWAs in [13, 17]. The cross-section of a singly-curved surface is usually an elongated region with curved boundaries, as shown in **Figure 22(a)**.

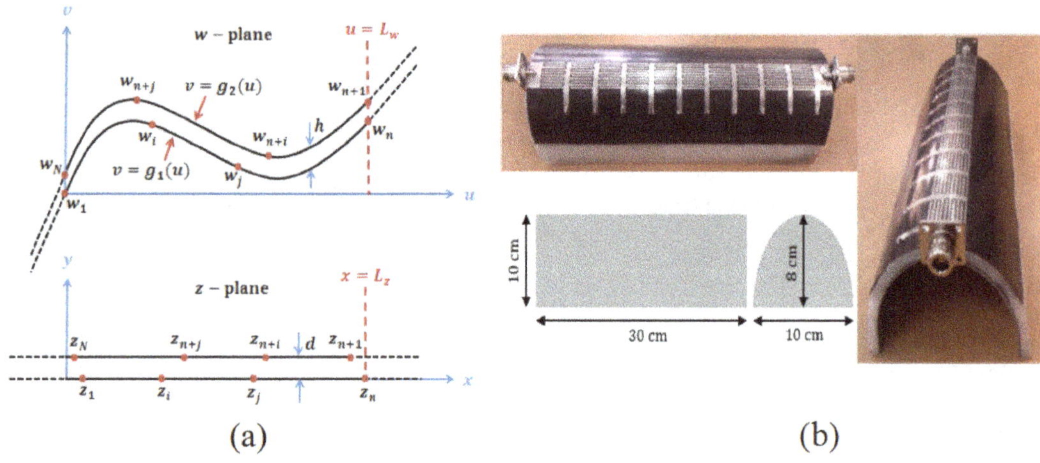

Figure 22. (a) Curved strip and the infinite straight strip mapped to the curve and (b) fabricated conformal CRLH LWA [13].

An elongated polygon may be mapped through the Schwartz-Christoffel transformation (SCT) to a new finite polygon. However, this may result in an undesirable effect called crowding. In order not to let the crowding effect take place, the transformation is adapted to map a long and straight strip to an infinite polygon. This polygon must have two ends at infinity. This is shown in **Figure 22(a)**. Curvature is accounted for by fitting each segment of the curved boundaries with a second degree polynomial function of SCT prevertices. This mapping is called modified Schwarz-Christoffel transformation (MSCT).

In some applications the utilization of conformal antennas is inevitable. For example on a vehicles' body if the antenna does not completely stick to the surface, it causes trouble in aerodynamics point of view. Hence, a conformal antenna which also makes the vehicle less visible, is installed on the body. In the reference [13], the MSCT approach is used to design a CRLH LWA on an elliptic surface, as shown in **Figure 22(b)**. The dispersion diagram and Bloch Impedance of a unit cell of the antenna are shown in **Figure 23**, which show good agreement between the MSCT method and full wave simulation (CST commercial package). Bloch impedance is a counterpart to characteristic impedance in periodic TLs. Also, the theoretical results predicted the measurements of the CRLH LWA.

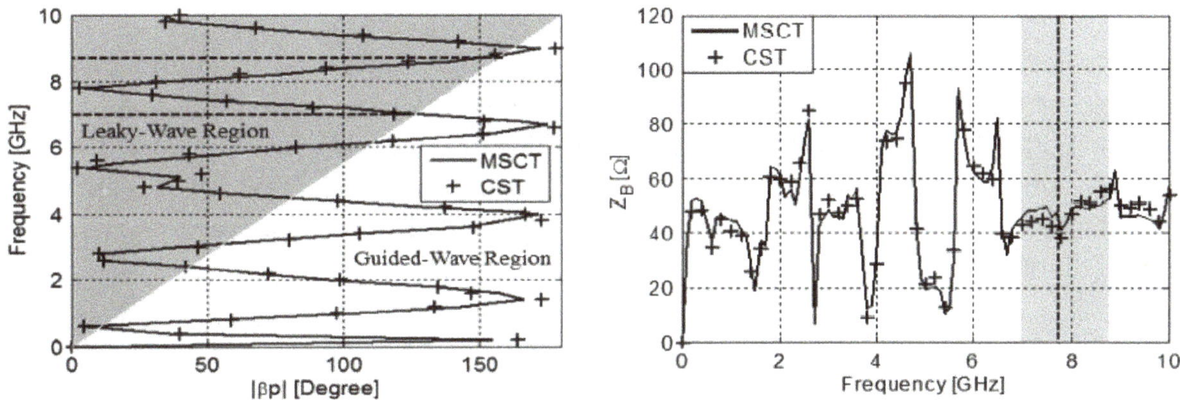

Figure 23. The dispersion diagram and Bloch impedance of the elliptic CRLH LWA unit cell. The shaded region is attributed to the leakage radiation frequency band [13].

10. Broadband CRLH LWA designed on dual-layer SIW [18]

In the reference [18], a dual-layer SIW X-band LWA with a bandwidth of 66 % beside 105° of the beam steering angle is obtained both in RH and LH regions. The top and side views of the wideband antenna are shown in **Figure 24**. The radiation aperture is provided by the interdi-

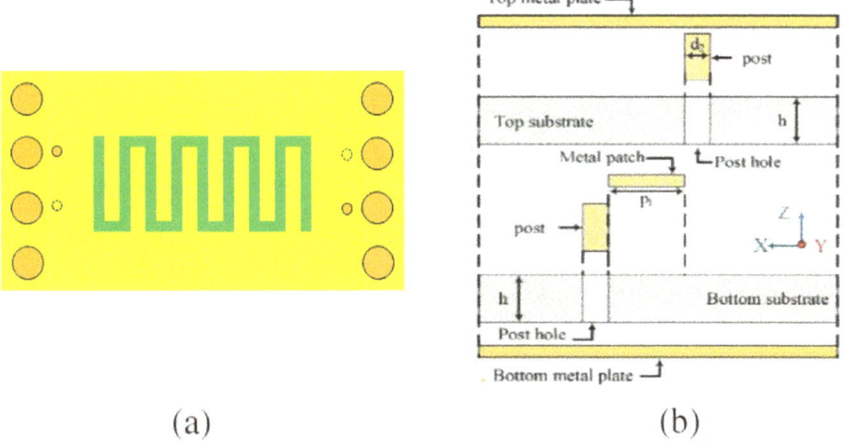

Figure 24. The designed dual-layer SIW antenna, (a) top view and (b) side view [18].

gital capacitor slots, as shown in **Figure 24(a)**. Increasing the total inductance of the structure is a way to tackle the existence of band gap at the operational bandwidth of the antenna. In this way the balanced conditions will be satisfied in a higher frequency, so the LH region will be extended. Using inductive elements to increase the inductance of the structure much more than what is provided by the nature of waveguide, is obtained with a twisted post which is realized by two metallic vias in the top and bottom substrates and are connected to each other using a metal patch etched on the top surface of the bottom substrate. **Figure 24(b)** illustrates the metallic posts and the metal patch in order to obtain the twisted post. The side wall vias of the SIW are not shown.

The dispersion diagram of the proposed unit cell in [18] is depicted in **Figure 25(a)**. The balanced condition is satisfied at 6.8 GHz. A wide frequency band attributes to the radiating region in both LH and RH bands which yields a wideband LWA. The scattering parameters of the LWA consisting of 10 unit cells are depicted in **Figure 25(b)**. S_{11} is less than − 10 dB and S_{21} is kept below −3 dB, which is suitable for operating the LWA. The antenna provides a fractional bandwidth of 66%.

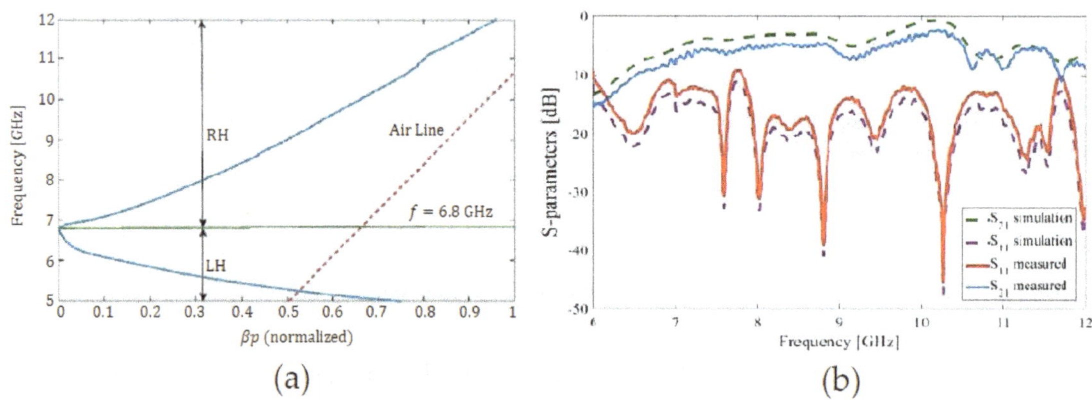

Figure 25. (a) Dispersion diagram of the unit cell and (b) scattering parameters of the 10-cell LWA [18].

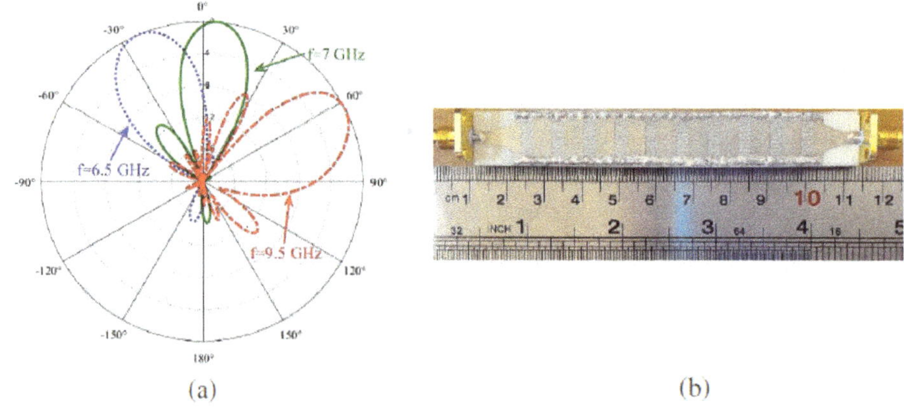

Figure 26. (a) The radiation patterns and frequency scanning of the beam and (b) the fabricated wideband dual-layer SIW LWA [18].

The radiation patterns of the X-band antenna with the property of beam sweeping are shown in **Figure 26(a)**. The LWA can change the direction of main lobe from −34° up to +72° including the broadside. There is a tradeoff between the bandwidth and the gain of the antenna. The sensitivity of gain, side-lobe level (SLL) and half-power beamwidth (HPBW) of the antenna in [18] is low and this is an important feature.

The fabricated antenna with the total dimensions of 100.2 mm × 13.53 mm × 1.69 mm is shown in **Figure 26(b)**. A wide variety of techniques are available to feed SIW structures, one of them is the tapering transition from microstrip to SIW, which is used in [18] to feed the LWA.

11. Single radiator CRLH circularly polarized LWA [19]

In the reference [19], a compact SIW CRLH-loaded LWA is proposed. The main advantage of the antenna is its pure circular polarization and also its compact size. A schematic of the unit cell is demonstrated in **Figure 27(a)**. Two interdigital slots are embedded in a single cell. The sections 2 and 3 each make 90° phase difference. The interdigital slots have angular separations of +45° and −45° with respect to the direction of wave propagating in the waveguide. This generates a circularly-polarized leaky wave radiation. As shown in **Figure 27(a)**, two microstrip lines are added to the cell in sections 1 and 4, with each section making 90° phase shift. This introduces a total phase shift of 360° to the unit cell. Hence, all the unit cells in the LWA have in-phase radiation and the constructive effect improves the directivity of the LWA.

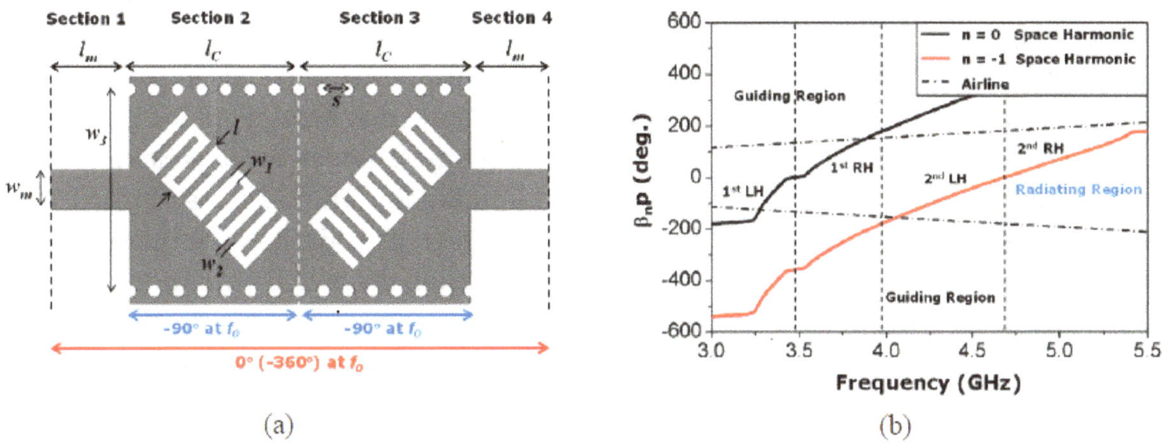

Figure 27. SIW circularly polarized CRLH LWA, (a) unit cell and (b) dispersion diagram [19].

The microstrip lines play the role of a matching circuit and this prohibits the use of an external matching network which leads to a compact antenna. Also, the placement of the two slots needs only one feed and hence there is no need to use a quadrature hybrid coupler to introduce a circular polarization. Therefore the antenna is a single radiator LWA. **Figure 27(b)** shows the dispersion diagram for the harmonics $n = 0$ and $n = -1$ which is clearly a multiband operation. The balanced conditions are also satisfied leading to the possibility of broadside radiation.

Figure 28(a) demonstrates the fabricated antenna and **Figure 28(b)** shows the LWA axial ratio clearly illustrating its relatively pure circular polarization. The application of circular polarization is inevitable in harsh environments. As an example a satellite moves dynamically and it is not possible to align the receivers (such as GPS system) to the satellite. Hence in this case the circular polarization is used instead of linear polarization.

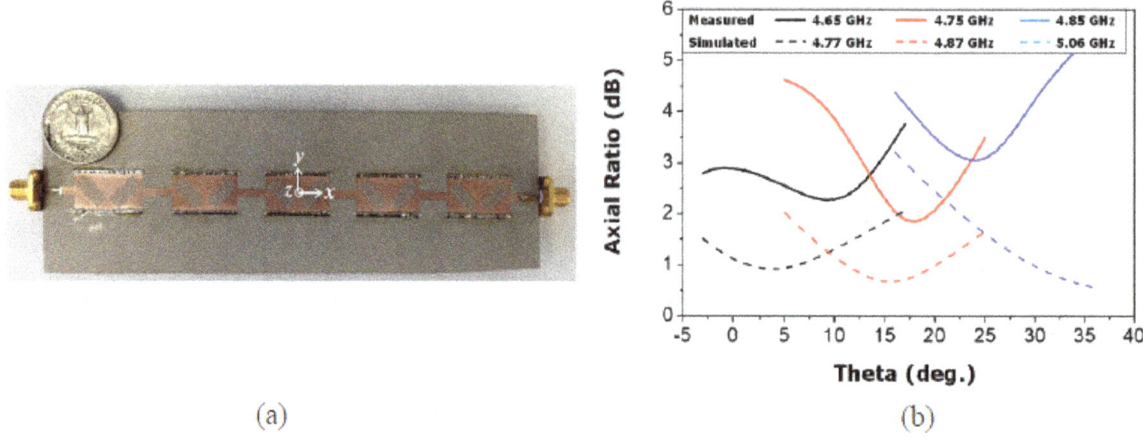

Figure 28. (a) The fabricated single radiator LWA and (b) the axial ratio diagram representing circular polarization [19].

12. Non-cutoff SIW-based LWAs [20]

In the reference [20] a SIW LWA is proposed which is based on a waveguide operating above its cutoff frequency. Note that in many conventional LWAs, the leaky wave radiation is based on below-cutoff guiding operation. A schematic of the proposed unit cells in [20] is demonstrated in **Figure 29**, where the left and right cells pertain to initial and modified unit cells. In the initial cell design in **Figure 29(a)**, the transverse slot can be modeled by a series impedance and the longitudinal slot is modeled by a shunt admittance. The field distribution in the transverse slot is minimum at its center and is maximum at its ends, hence it behaves like an inductor in the transition frequency. In the longitudinal slot, the field distribution is reversed and hence it plays the role of a capacitor. At the transition frequency, one can easily satisfy the condition $S_{11} = 0$ and therefore the open stopband can be suppressed. This results in a seamless broadside radiation.

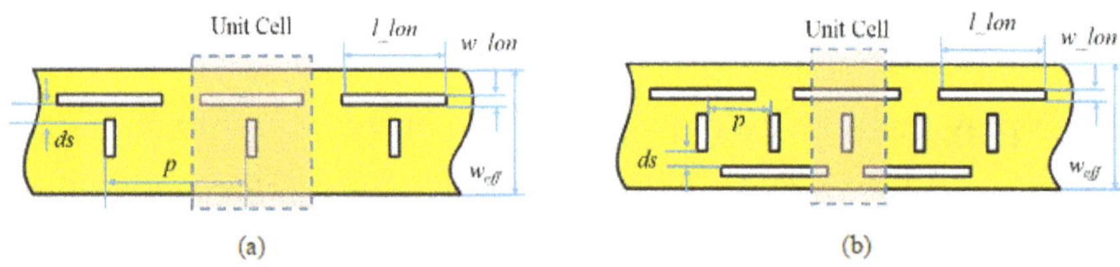

Figure 29. The unit cells for above-cutoff operation and suppression of open stopband in a LWA, (a) initial cell and (b) modified cell [20].

The scattering parameters for the simulated LWA are shown in **Figure 30(a)**. The transition frequency is 11.8 GHz and hence as shown in Figure REF fig30 \ h 30, the antenna can radiate in backward and forward angles. There is a peak in S_{11} at the transition frequency and this shows that even though the open stopband is suppressed, the effect of spurious reflections is observed. **Figure 30(b)** shows the main lobe direction and realized gain for the simulated antenna. The antenna supports scanning angles between −70° and 30°. The maximum realized gain happens at forward angles and is about 70 dB.

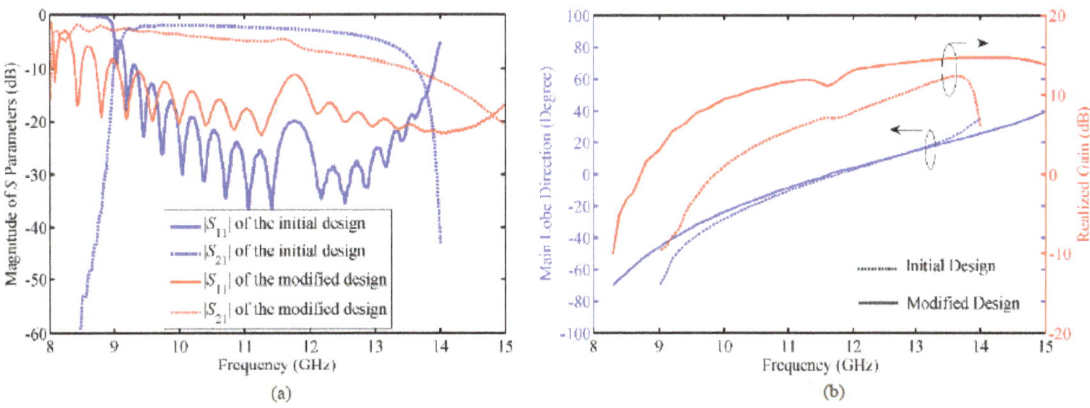

Figure 30. (a) Scattering parameters and (b) main lobe direction and realized gain [20].

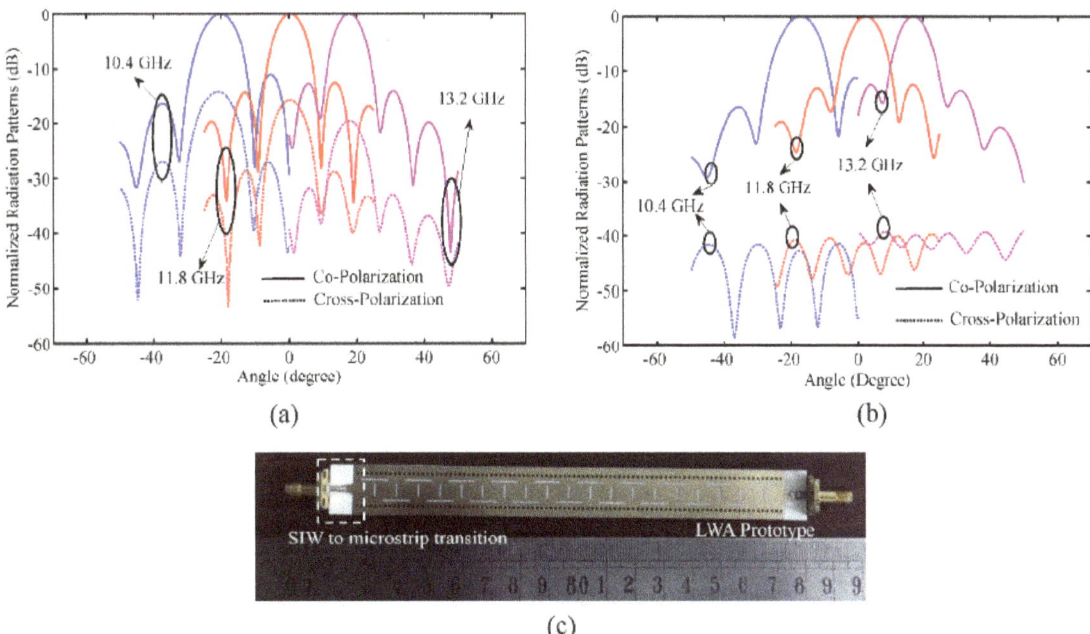

Figure 31. Backward, broadside and forward radiation patterns for the LWA comprising (a) initial unit cells and (b) modified unit cells. (c) The fabricated LWA using modified cells [20].

The radiation patterns for backward, broadside and forward angles for co- and cross-polarizations in the LWA with initial unit cells are shown in **Figure 31(a)**. The frequency scanning of the beam is clearly observed. When using the modified unit cell of **Figure 29(b)**, the antenna become more compact, since the condition of in-phase radiation for the cells is satisfied for a smaller unit cell. Also, as shown in **Figure 30(a)** it has a more efficient behavior in scattering parameters. It renders S_{11} below −10 dB and concurrently, reduces the insertion loss resulting in increasing the leakage rate. Accordingly, the realized gain is improved, as demonstrated in **Figure 30(b)**. Also, its gain is more flat compared to the initial unit cell. **Figure 31(b)** shows that the modified unit cell results in lower cross-polarization compared to **Figure 31(a)**. The fabricated antenna including modified unit cell is shown in **Figure 31(c)**.

13. Programmable LWA with J-shaped MTM patches [21]

Reference [21] has introduced a novel J-shaped unit cell which has the ability of digital tuning of the beam at a fixed frequency. This can be a potential alternative to complicated phased arrays. A schematic of the antenna with its unit cell is shown in **Figure 32(a)**. The J-shaped patch is a CRLH cell, since the shunt inductor is formed by the virtual ground at the top of the patch. Other circuit equivalents are clearly deduced. In each unit cell, the upper gap can be open (ON) or close (OFF). This is the base for digital tuning and is demonstrated in **Figure 32(b)**.

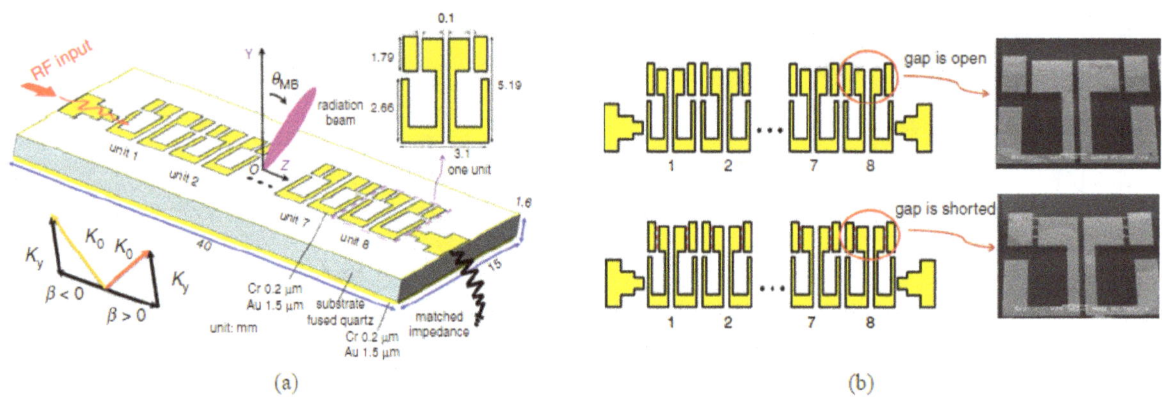

Figure 32. (a) Programmable LWA and its J-shaped patch unit cell and (b) the ON and OFF states in the J-shaped patch [21].

In a LWA with N unit cells, 2^N different digital states can be programmed. In the reference [21] a LWA with eight elements is considered and then 256 states are possible. Among the states, nine samples including different ON-OFF states are studied. Since at a fixed frequency, changing each state changes the propagation constant of the LWA, it also changes the main beam direction. **Figure 33(a)** shows the realized micromachined LWA and **Figure 33(b)** illustrates the beam scanning with digital states. The left and right patterns pertain to simulation and measurement results, respectively.

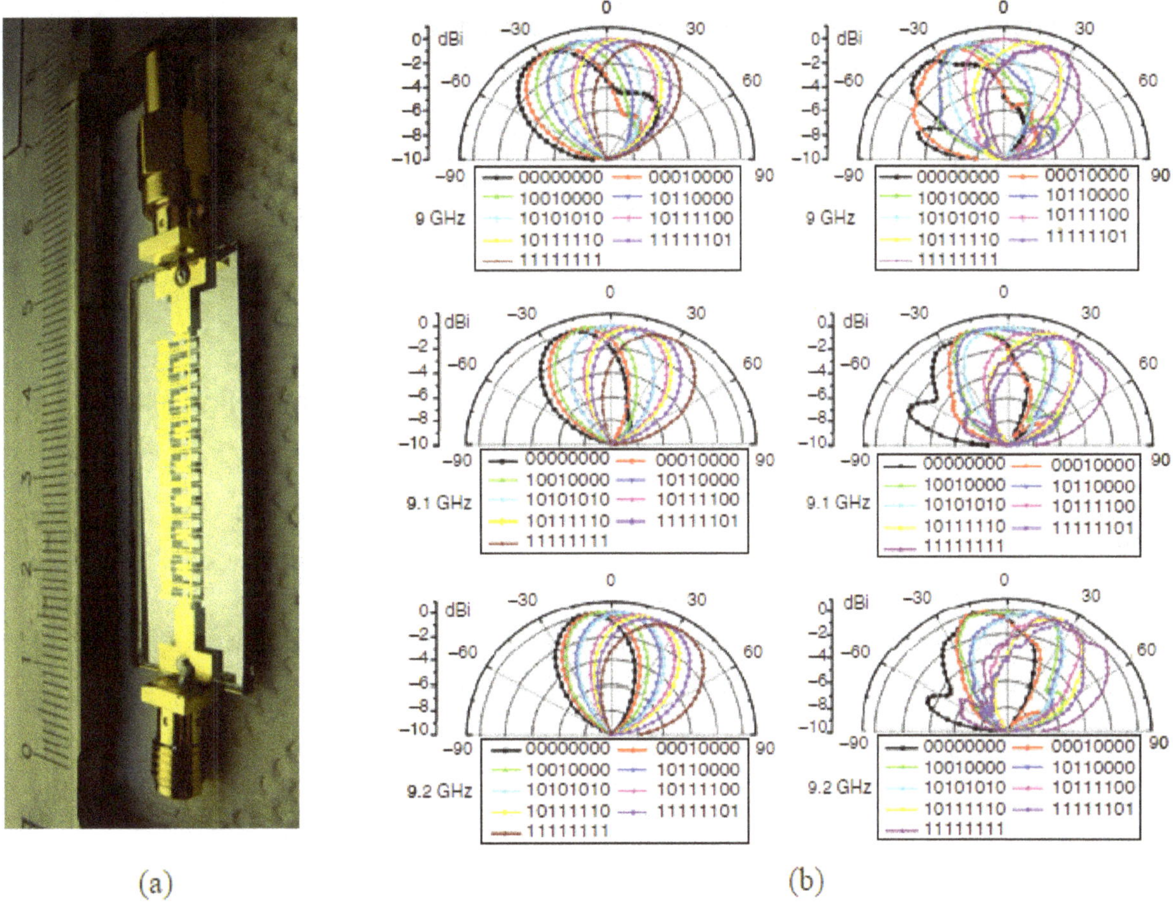

Figure 33. (a) Fabricated micromachined programmable LWA and (b) simulated and measured radiation patterns scanned with different digital states [21].

14. Conclusion

In this chapter the most significant works pertaining to CRLH LWAs have been reviewed. The most practical technology suitable for LWAs is a combination of microstrip and SIW which is compatible with integrated circuits. Also, SIW is a low loss technology which is able to carry high power. Four major beam scanning approaches were introduced which include frequency, bias voltage of a varactor, magnetic bias of a ferrite, and short/open gaps (digital states). The main characteristics of a CRLH unit cell are its dispersion diagram and Bloch impedance. The equivalent circuit of a CRLH cell comprises two capacitors and two inductors and can be implemented in many forms studied in the chapter. CRLH unit cells are not the only choice for LWAs but they have acquired a wide range of application and diversity. Leaky wave radiation is obtained based on either below-cutoff or the preferable above-cutoff waveguide operation. The LWAs can be made dual band, wideband, circularly polarized and reconfigurable. They are useful in reflecto-directive systems and can be integrated with active devices. The LWAs are promising alternatives to phase shifters which greatly reduce the cost and complexity of the beam scanning systems. Specifically they can be conformed to the body of satellites where a simple light-weight technology becomes crucial.

Author details

Keyhan Hosseini* and Zahra Atlasbaf

*Address all correspondence to: k.hosseini@modares.ac.ir

Faculty of Electrical and Computer Engineering, Tarbiat Modares University, Tehran, Iran

References

[1] A. Oliner. Leaky Wave Antennas. In: Antennas Engineering Handbook. 3rd. ed. New York: McGraw Hill; 1993.

[2] J. D. Kraus, and R. J. Marhefka. Antennas. New York: McGraw Hill. 2001.

[3] A. Ishimaru. Electromagnetic Wave Propagation, Radiation, and Scattering. New Jersey: Prentice Hall; 1991.

[4] C. Caloz, and T. Itoh. Electromagnetic Metamaterials: Transmission Line Theory and Microwave Applications. Hoboken, NJ: Wiley intersci.; 2006.

[5] A. Sanada, C. Caloz, and T. Itoh. Characteristics of the composite right/left-handed transmission lines. IEEE Microwave and Wireless Components Letters. 2004;14(2):68–70.

[6] S. Lim, C. Caloz, and T. Itoh. Metamaterial-based electronically controlled transmission-line structure as a novel leaky-wave antenna with tunable radiation angle and beamwidth. IEEE Transactions on Microwave Theory and Techniques. 2005;53(1):161–173.

[7] Y. Dong, and T. Itoh. Composite right/left-handed substrate integrated waveguide and half mode substrate integrated waveguide leaky-wave structures. IEEE Transactions on Antennas and Propagation. 2011;59(3):767–775.

[8] F. P. Miranda, C. Penalosa, and C. Caloz. High-gain active composite right/left-handed leaky-wave antenna. IEEE Transactions on Antennas and Propagation. 2006;54(8):2292–2300.

[9] S. Lim, C. Caloz, and T. Itoh. A reflecto-directive system using a composite right/left-handed (CRLH) leaky-wave antenna and heterodyne mixing. IEEE Microwave and Wireless Components Letters. 2004;14(4):183–185.

[10] J. Machac, M. Polivka, and K. Zemlyakov. A dual band leaky wave antenna on a CRLH substrate integrated waveguide. IEEE Transactions on Antennas and Propagation. 2013;61(7):3876–3879.

[11] T. Kodera, and C. Caloz. Dual-band full-space scanning leaky-wave antenna based on ferrite-loaded open waveguide. IEEE Antennas and Wireless Propagation Letters. 2009;8:1202–1205.

[12] G. Zamora, S. Zuffanelli, F. Paredes, F. J. H. Martinez, F. Martin, and J. Bonache. Fundamental-mode leaky-wave antenna (LWA) using slotline and split-ring-resonator (SRR)-based metamaterials. IEEE Antennas and Wireless Propagation Letters. 2013;12:1424–1427;

[13] K. Hosseini, and Z. Atlasbaf. Analysis and synthesis of singly-curved microstrip structures utilizing modified Schwarz-Christoffel transformation. IEEE Transactions on Antennas and Propagation. 2013;61(12):5940–5947.

[14] K. Hosseini, and Z. Atlasbaf. Mutual coupling between two microstrip patch antennas on concave and convex cylindrical surfaces. In: Proceedings of the IEEE National Conference on Electrical Engineering (ICEE2013); 1–4 May 2013.

[15] K. Hosseini, and Z. Atlasbaf. Guided- and radiated-wave characteristics of a rectangular patch antenna located on a singly-curved surface. Progress in Electromagnetic Research C. 2013;38:205–216.

[16] K. Hosseini, and Z. Atlasbaf. Mutual coupling between two microstrip patch antennas on a singly curved surface. IEEE Antennas and Wireless Propagation Letters. 2013;12: 313–316.

[17] K. Hosseini, and Z. Atlasbaf. Design of a cylindrical CRLH leaky-wave antenna using conformal mapping. In: Proceedings of the IEEE International Symposium on Telecommunications (IST2012); 33–36 November 2012.

[18] M. Niayesh, and Z. Atlasbaf. Broadband CRLH beam-scanning LWA designed on dual-layer SIW. Applied Computational Electromagnetics Society Journal. 2016;31(4): 450–454.

[19] H. Lee, J. Choi, C. M. Wu, and T. Itoh. A compact single radiator CRLH-inspired circularly polarized leaky-wave antenna based on substrate integrated waveguide. IEEE Transactions on Antennas and Propagation. 2015;63(10): 4566–4572.

[20] Y. L. Lyn, X. X. Liu, P. Y. Wang, D. Erni, Q. Wu, C. Wang, N. Y. Kim, and F. Y. Meng. Leaky-wave antennas based on non-cutoff substrate integrated waveguide supporting beam scanning from backward to forward. IEEE Transactions on Antennas and Propagation. 2016;64(6):2155–2164.

[21] Y. Lu, K. Kikuta, Z. Han, T. Takahashi, A. Hirose, and H. Toshiyoshi. Programmable leaky-wave antenna with periodic J-shaped metamaterial patches. Electronics Letters. 2015;51(10):733–734.

6

Micro Switch Design and its Optimization Using Pattern Search Algorithm for Application in Reconfigurable Antenna

Paras Chawla and Rohit Anand

Abstract

This chapter reports the design and optimization algorithm of metal-contact RF microswitch. Various important evolutionary optimization techniques that can be used to optimize non-linear and even non-differentiable types of radio frequency (RF) circuit's problems are also reviewed. The transient response of the proposed switch shows displacement time (i.e., squeezed-film damping effect) of 5.0 μs and pull-in voltage varying from 9.0 to 9.25 V. Primarily, the switch exhibits insertion loss of 0.15 to 0.51 dB in on-position and isolation of 75.96 to 35.83 dB in off-position at 0.1–10 GHz. Also, the proposed RF switch equivalent circuit and layout are validated in ADS software which was earlier simulated in HFSS. A pattern search (PS) algorithm is used to optimize RF characteristics of the proposed switch after a brief review of the different optimization techniques. After optimization, the switch shows decrement in insertion loss and increment in isolation at 0.1–10 GHz. Further, two such optimized switches are introduced on the defected ground structure (DGS) antenna to make it reconfigurable in terms of frequency. Reconfigurable antenna (RA) is simulated using HFSS software and simulation results are verified by showing the mark of agreement with the fabrication results. The novelty in the proposed design is due to dual-band behavior and better resonance performance than antennas available in the literature. Attractions of proposed RA are its miniaturization and its utility in IEEE US S-(2.0–4.0 GHz) and C-(4.0–8.0 GHz) band.

Keywords: design optimization, microstrip antennas, RF microswitches, reconfigurable antenna, vehicular and wireless technologies

1. Introduction

There are various kinds of RF switches available that can be used along with microstrip antenna to achieve the reconfigurability. An RF MEMS metal-contact switch is an emerging technology that has replaced the semiconductor field-effect transistors (such as GaAs FETs), PIN diode switches/Schottky diodes and electromagnetic relays [1, 2]. The metal-contact MEMS switches are not only used in reconfigurable antenna structures but also in switching networks, satellite systems, filters and automated test equipments from 0.1 to 40 GHz applications [3].

MEMS RF switches demonstrate exceptional performance in upper range of frequency as compared to the traditional semiconductor RF switching microelectronics technology. The advantages of MEMS RF switches over semiconductor switches are batch fabrication (that involves lithography-based micromachining, fabricated on either quartz or automated mark high-resistivity Si or GaAs wafers), low insertion loss (around 0.2 dB), high isolation (around 70 dB), small off-state capacitances (2–4 femto Farads), high linearity and low power consumption (almost zero power). Also, metal semiconductor (MES)-FET switching in the cold-FET mode requires almost no control power. Some commercial cold-FETs can switch at 2.3 V [4–7].

Performance comparison of the various commercially available RF switches is shown in **Table 1**.

Despite all of the advantages discussed, MEMS switches also have some problems associated with them. The main issues at present are relatively low speed (around 3–40 μs), moderate voltage or high current initiative (electrostatic actuated switches require 5–80 V for consistent operation, whereas magnetic/thermal switches can be activated with 2–6 V but require 9–105 mA of current supply), reliability (0.2–4 billion cycles), high packaging cost and low power handling (<200 mW) and therefore, they are rarely commercially available [4, 7].

Optimization of MEMS switches had played a major impact in RF circuits [10–12]. The microstrip patch antennas (MSA) are preferred for wireless applications due to their low-profile

Parameter of switch characteristics	MEMS	GaAs/metal semiconductor (MES)—field-effect transistor (FET)	PIN-diode	Electromechanical relay (EMR)
Voltage (V)	10–75	3–5; 2.3 (commercial cold-FETs)	±2.5–5	3–2
Current (mA)	Almost zero	Almost zero	0–20	15–150
Power consumption (mW)	0.05–0.1 (negligible power consumption)	0.05–0.1; no power consumption in commercial cold-FETs	5–95	<380
Switching time	1–400 μs	1–120 ns	10–120 ns	>1 ms
Isolation in OFF condition (1–10 GHz) (in dB)	>45	15–25	>35	>40
Insertion loss in ON condition (1–10 GHz) (in dB)	0.05–0.2	0.4–2.5	0.3–1.2	<0.3

Table 1. Comparison of various electrostatic RF switches [4, 8, 9].

structure, easy to manufacture structure, wide and multiband behavior [13, 14]. Different techniques have been used to achieve multi-band operation for MSA. Some of the techniques employed variation in physical dimensions of feed line, modification of the effective length of antenna using slot, implementation of defected ground structure and use of the switching device within the antennas [15–17].

Further, the fabrication process of the RA is a difficult multi-layered task and an effective integration needs cautious planning for every single stage in the direction not to fade the proposed antenna's performance. Among those stages, the utmost critical procedures [4, 13] can be summarized in:

- *Release and fabrication procedure of the RF switch membrane.* Proper care must be required for this step; otherwise, stiction complications, membrane deformations and infatuation of sacrificial layer occur.
- *Patterning of DC bias transmission line.* It requires very well contact by means of the DC pad; otherwise, there are chances of leakage of RF energy through transmission line.
- *Patterning and the deposition of the thin film of dielectric material designed for the RF MEMS switches.* Imprecision all through this stage may cause either the major dielectric charge issues or the small break down voltage of the designing switches.

All the aforementioned issues can be avoided if the procedure is well organized; otherwise, it will affect the electromagnetic characteristics as well as input impedance of the antenna [4, 13].

The motivation of this research work is a great effort put in emerging RF-switch-based-reconfigurable antenna designs and procedures capable of facing specific challenges such as low power consumption, simplicity, robustness as well as small size. The scope of this work describes a two-way application: (a) design of an electrostatic actuated metal-contact cantilever beam switch and (b) development of a grid-based numerical optimization method based on pattern search (PS) technique. Further, some important nature-inspired-optimization algorithms and numerical-based-optimization algorithms are also studied that can be helpful to improve complex RF circuits like switches and antennas. The design of cantilever beam microswitch is identified for applications in microwave multiband reconfigurable antennas. A novel approach of PS optimization method is used here to optimize the isolation, insertion and return loss of metal-contact microswitch. Finally, all the above-mentioned procedure is combined together that creates a slot-dual frequency band reconfigurable antenna configuration.

2. Review of some important optimization techniques

Optimization refers to maximizing or minimizing the objective function or fitness function. In today's scenario, the evolutionary optimization techniques that can be used to optimize even non-differentiable and non-linear types of problems are most commonly used. These techniques are based on intensifying the search (w.r.t. neighborhood) and diversifying the range (of solutions). Some of these significant algorithms will be discussed in this section:

Particle Swarm Optimization (PSO): PSO [18, 19] is a population-based and swarm-intelligence-based evolutionary optimization technique motivated by the flocking of the birds and the schooling of the fishes. The solution of the problem is represented as a particle and the parameters of the solution are represented as 'position, velocity, local best position and global best position' of the particle. All the positions achieved are evaluated by a kind of fitness function to represent how well the design criterion (e.g., switch design or antenna design criterion) is satisfied. The process terminates after a pre-specified number of iterations. This technique is very simple and fast because of being based on intelligence and learning. But the quality of results may not be so good because of being based upon a random process.

Genetic Algorithm (GA): GA [19] is a population-based evolutionary optimization technique motivated by the natural evolution process. The solution of the problem is represented as a chromosome and the parameters of the solution are represented as genes. The problem is initialized by the population of chromosomes. The best chromosomes are selected (that act as parents) using an appropriate fitness function (for example, in switch design, the solution parameter may be the insertion loss or isolation as in our case). These parents then generate the offspring using 'crossover' and 'mutation'. Last step is elitist replacement scheme that compares all the individuals in the population and the offspring. The program terminates after a designated number of iterations. It can be applied to a wide range of simple as well as complex problems. But it may be time-consuming sometimes to find an optimum solution.

Ant Colony Optimization (ACO): ACO [20] is another population-based and swarm-intelligence-based evolutionary optimization technique motivated by the foraging behavior of the social ants. The solution of the problem is represented as a pheromone deposited by the ants in going toward and back from the food source and the parameters of the solutions are represented as the concentration of the pheromone. The other ants follow the route at which pheromone concentration is higher. The food may be called as the desired condition, for example, in switch optimization using ACO, some particular range of the insertion loss or isolation may be the desired conditions. This technique is very fast and efficient because of being based on intelligence and learning. But sometimes it yields the local optimum solution because this technique updates the pheromone according to the current best path.

Firefly Algorithm (FA): FA [21, 22] is one more population-based and swarm-intelligence-based evolutionary optimization technique motivated by the short and rhythmic flashing patterns of the swarming fireflies. Fireflies are the glowworms that can produce the natural light (with its intensity proportional to its hunger) to attract victim. For example, in switch design context, this technique may be used to compute the optimum weights and positions of the switches for the optimum design. This technique converges very fast and it is based on intelligence and learning. But there is no provision of the memory to memorize the situation that remained better than the present.

Cuckoo Search (CS): CS [21, 23] is another population-based and swarm-intelligence-based evolutionary optimization technique motivated by the hatching behavior of cuckoo species. The female cuckoo chooses a nest to lay the eggs where the host has just laid its own eggs. In this technique, each egg in the nest corresponds to a solution while each cuckoo egg corresponds to a new solution. Keeping the best nest corresponds to the best objective. This quality optimization technique may be used to optimize many of the parameters in switch design. It is a simple, robust and easy to converge method. But the parameters of this approach cannot be changed that results into less efficiency.

Bat Algorithm (BA): BA [21, 24] is one more population-based and swarm-intelligence-based evolutionary optimization technique inspired by the echolocation nature of the microbats. They use echolocation to detect their prey even in the whole darkness, that is, they transmit a sound pulse and listen to the echo bouncing back from the surrounding objects. For example, locations of the switches in a reconfigurable antenna may be optimized using this technique. This technique becomes more optimum with the increase in population size and it is based on intelligence and learning. But it has somewhat low convergence accuracy.

Artificial Bee Colony Optimization (ABC Optimization): ABC algorithm [25] is another popular population-based and swarm-intelligence-based evolutionary optimization technique inspired by the behavior of honeybee in food foraging. In this technique, solution to the optimization problem is indicated by the position of a food source while the solution quality (or fitness) is indicated by the food source. Each employed bee corresponds to one food source (i.e., one solution). An onlooker bee selects source of food depending on the probability value associated with the food source. If a food source (i.e., solution) cannot be improved further, the scout bee helps to generate new solutions randomly. For example, this algorithm may be effectively used to tag the switch design problems. It provides the clarity and errors in case of optimal solutions, but it has not been so widely used for solving the real-life problems.

Galaxy-based Search Algorithm (GbSA): GbSA [26] is a new evolutionary algorithm that has come into existence only a few years back. In this technique, the movement of solution is spiral from randomly generated initial solution (initial solution is commonly assumed to be at the core of galaxy) and the arm of galaxy moves spirally to search the surrounding until it finds a better solution. The algorithm is very optimal, but it needs to be used for many of the applications in the future.

Harmony Search (HS): HS [27, 28] is a population-based evolutionary optimization algorithm (mainly used for solving the reliability problems) motivated by the nature of music. This method may convert the harmony of music into optimization. Just like music in which a particular note is played to have the pleasing harmony, a particular value is created in each decision variable to have the best possible optimum value. Like the judgement of quality of music by the pitch function, judgement of the quality of the decision variable is done by the fitness function. This technique does not need any initial value for the optimization, but it performs well only for a single objective function.

Biogeography-Based Optimization (BBO): BBO [29] is a population-based evolutionary algorithm working on migration and mutation. The algorithm is based on the principles of emigration and immigration between the habitats. A habitat with high value of habitat suitability index (HSI) has high emigration rate and low immigration rate and vice versa. In this technique, each habitat corresponds to a solution, that is, high value HSI habitat corresponds to a good quality solution and vice versa. As an example, BBO technique may be used for the optimization of post-parameters in antenna design or switch design. This method is fast and free from any assumptions. Moreover, good solutions are retained. But the system using only migration and mutation may not converge to the global optimum.

Differential Evolution (DE): DE [30] is another population-based evolutionary computational algorithm for the optimization. It uses a differential operator to create a new solution. This algorithm has been widely used in the design and synthesis of antenna as well as switches. The main difference between DE and GA is the use of same operators but in the different ways. In this method, the selection step is implemented after the mutation and crossover steps and involves both the parents as well as offspring. This method is good in diversification, but it has somewhat less accuracy.

Simulated Annealing (SE): SE [31] is an evolutionary probabilistic optimization technique motivated by the annealing process in solids. It is a technique without any memory. This algorithm is based on the trajectory of the search path. In this technique, a material is heated above melting point and then is cooled gradually to the ambient temperature to generate a larger crystalline solid with minimum energy probability distribution and minimum metallic defects. Most of the times, this is done in a lot of iterations. This technique can efficiently be used for the global optimization of the different parameters in switch design. It is good even for quite unordered data and it has very good global optimality. But there is a trade-off between the computational speed and quality of the solution.

Invasive Weed Optimization (IWO): IWO [32] is an evolutionary optimization technique inspired by the colonization of the invasive weeds. Each invading weed grows to a flowering weed and generates new weeds. These new weeds again grow to the flowering weeds and this process goes on and on and at last, maximum number of weeds is spread over the field. Now, only those weeds having good fitness can survive to generate new weeds. Each individual (i.e., a set having a value of each optimizing variable) is represented by a seed while any individual after evaluating its fitness is represented by a plant. As an example, this technique can be applied easily for the reconfigurable antennas by the positions of the switches. This technique shows a very high stability, but there is a further scope of improvement in the rate of success of this technique.

Pattern Search (PSearch or PS) Optimization: PSearch (or PS) Optimization [10, 33] is a very efficient, non-random and direct (i.e., does not require any information about the gradient of the fitness or objective function) optimization tool (especially with regard to multi-objective optimization in antenna design or microswitch design) for searching the minima of a function which is not necessarily differentiable or even continuous. This optimization strategy is comparatively faster than the other discussed algorithms. The algorithm searches a collection of points (called mesh having some specified pattern and constant size) surrounding the current point until a point is found in that mesh where the fitness or objective function value is less than its value (in case of minimization) at the current point. Thereafter, this new point acts as current point and once again starts searching the neighborhood points to obtain a new optimized point in the next iteration of the algorithm. If there is no point on mesh where the objective function value is lower than its value at the current point, then this

Feature → technique ↓	Inspired by	Time for convergence/ optimization	Accuracy	Suitability
Particle swarm optimization [18, 34]	Flocking behavior of birds	Little bit more time (because of overhead)	Average	Even for complex problems
Genetic algorithm [19, 34]	Natural evolution	Little bit more time (because of overhead)	Good	Even for complex problems
Ant colony optimization [20]	Behavior of social ants	Less time	Average	Even for complex problems
Firefly algorithm [21, 22]	Flashing behavior of the fireflies	Less time	Good	Even for complex problems
Cuckoo search [21, 23]	Hatching behavior of cuckoo	Quite less time	Good	Even for complex problems
Bat algorithm [21, 24]	Echo behavior of bats	Little bit more time (because of being fallen into local optimization sometimes)	Average	Even for complex problems
Artificial Bee colony optimization [25, 35]	Behavior of honeybee in food foraging	Less time	Good	Even for complex problems
Harmony search [27, 28]	Nature of music	Less time	Average	For simple problems
Biogeography-based optimization [29, 36]	Distribution of plants and animals in habitats	Less time	Good	Even for complex problems
Differential evolution [30, 37]	Natural laws concerned with the evolution of the individuals	Less time	Average	Even for complex problems
Simulated annealing [31, 38]	Annealing process in solids	Higher time	Good	Even for complex problems
Invasive weed optimization [32, 39]	Colonization of the invasive weeds	Little high	Low	Even for complex problems
Pattern search optimization [10, 33]	Searching around mesh	Less time	Good	Even for complex problems

Table 2. Comparison of important optimization techniques.

algorithm will reduce the size of mesh so as to search around the current point with more persistence and hence to find the minimum with more accuracy.

A very brief comparison of these optimization techniques has been discussed in **Table 2** (more experiments need to be done for GbSA as it has not been used in so many cases).

A very brief review of the advantages and disadvantages of the different optimization techniques discussed above can be shown in **Table 3**.

In this research work, PSearch algorithm will be used for the synthesis and optimization of the switch because of its speed, reliability and accuracy. Although PSearch is a much faster method of reaching the acceptable results, but it can be improved more by selecting the different search directions depending upon the objective function to be minimized.

Optimization technique	Advantages	Disadvantages
Particle swarm optimization [18, 40]	1. Particles preserve the good solutions (memory characteristic) 2. Each member is involved in the exploitation of the solution space	1. Quality of results may not be too good as the guided random process is the basis 2. Very sensitive to the tuning of parameters
Genetic algorithm [38]	1. Very useful for the large or/and unknown search space 2. Solution becomes more and more optimal with passage of time	1. Solution is prone to fall in local optimum if the fitness (cost) function is not properly defined 2. Not a very good technique for the constraint-based optimization
Ant colony optimization [38]	1. Adaptable to real time changes 2. Has property of distributed computing	1. Convergence is difficult to prove in many cases 2. Solution may fall into local optimum in a few cases
Firefly algorithm [22]	1. High speed and performs good even at high levels of noise 2. Performs better with the increase in population size	1. Unable to get rid of local optima while doing local search (in a very few situations) 2. Unable to preserve the good solution (no memory)
Cuckoo search [23, 41]	1. Tendency to be trapped in local optimum is very very less and hence faster than the other methods 2. Easier to implement as it depends upon population and probability only	1. Little less efficiency because of inability to change parameters 2. Number of iterations required may be quite large
Bat algorithm [24, 42]	1. Very efficient even for the multi-constraint problems 2. Can be used to derive many algorithms by changing some of its parameters	1. Little less convergence accuracy 2. Not so good for the multi-dimensional problems because of fast initial convergence
Artificial Bee colony optimization [35]	1. Can intensify and diversify the search quite effectively 2. Gives optimum results in reasonable time	1. Is yet to prove its worth in so many real-life applications 2. Some mathematical justifications are yet to be done in this technique
Harmony search [43, 44]	1. Does not need any initial value resulting in convergence to a global optimum in a much better way 2. Parameters need not to be fine tuned	1. Mainly useful only for a single objective function 2. Very poor in case of local search
Biogeography-based optimization [45]	1. Has high diversity resulting in a good solution 2. Preserves the good solutions (memory characteristic)	1. Less efficient in exploitation of the solution 2. Sensitive to the tuning of some parameters
Differential evolution [30, 37]	1. Efficient memory utilization 2. Can easily be implemented in parallel	1. Large computation needed to reach the optimal solution 2. Less accurate
Simulated annealing [38]	1. Ability to approach the global optimal solution in most of the cases 2. Simple to code even for the complicated and unordered data/problems	1. Very much time-consuming because the fitness function needs more computation 2. Parameters have to be very much fine-tuned

Optimization technique	Advantages	Disadvantages
Invasive weed optimization [32, 39]	1. Allows all plants to take part in reproduction 2. Shows high stability and good convergence	1. Little scope of improvement in the success rate of this algorithm 2. Is yet to prove its importance in many real-life situations
Pattern search optimization [10, 33]	1. Very fast and reliable method 2. Very accurate method as the mesh size may be reduced to search around the point with higher resolution	1. Wrong selection of starting point may result in sticking around local minima

Table 3. Advantages and disadvantages of important optimization techniques.

3. Switch design and analysis

Initially, metal MEMS switch is analyzed with the help of software coventorware. RF MEMS switches in terms of multi-Physics properties have been studied, designed and simulated for RA design. The multi-Physics characteristics for MEMS switches generally include the squeeze film damping effect/displacement time, activation/pull-in voltage, electrostatic force and on-off capacitance ratio.

The lumped/equivalent circuits of aforementioned MEMS switch are generated by using Advanced Design System (ADS) software with the help of Hspice model. After generation of lumped elements, the mathematical analysis and verification of post-processing simulation results of switches are performed. Further, the ANSYS high frequency structure simulator (HFSS) software is used to find out the electromagnetic (EM) characteristics of MEMS switch. In case of switch, the important EM post-parameters like insertion loss (IL), return loss (RL) and isolation are studied. This software provides the in-built optimization tools which are also considered to optimize the EM results.

Proposed MEMS switch here moves at one end in the downward direction and is fixed at other end of the beam. The cantilever beam makes a metal contact with the transmission line as it moves in down direction in response of electrostatic actuation. When beam metal part connects the two ports of transmission line, switch is called in ON position and when beam metal part disconnects the two ports of transmission line, the switch is called in OFF position [46]. The presented MEMS resistive switches are different according to their beam shapes and metal contact areas.

Figure 1 shows 3D view of the metal-contact RF microswitch implemented on silicon substrate of area 272 × 118 μm^2 and thickness 48 μm. The cantilever beam is made of gold and silicon nitrate of thickness 0.2 μm. A 0.5-μm-thick dimple is used to make metal to metal contact and to separate the cantilever beam from actuation pad in down-position.

The equivalent R-L-C circuit of proposed switch in on-position and off-position are shown in **Figure 2**. The circuit model of metal-contact microswitch is used to extract the *C, R* and *L* equivalent parameters of the switch. There are a number of circuit elements which represent the physical part of the proposed microswitch [47]. Here in **Figure 2a**, R_4 = 3.98 ohm represents a value of line resistor; L_2 = 0.63 nH, a value of line inductor of the cantilever

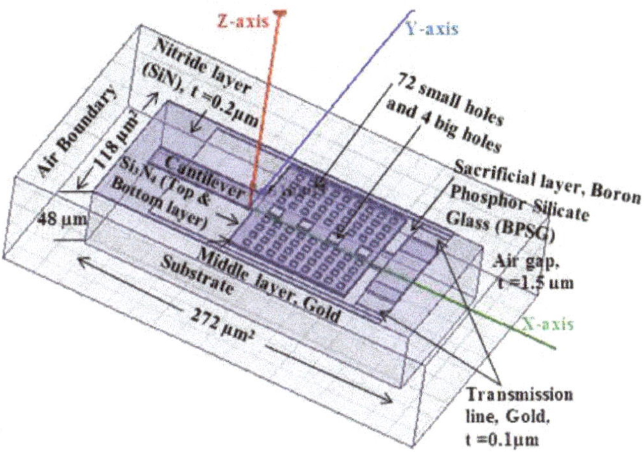

Figure 1. 3D layout of metal contact microswitch (in coventorware software).

beam. Further, R_3 = 3.3 ohm, contact resistance value; L_1 = 1.35 nH, contact inductance value; and C_1 = C_2 = 0.017 pF, the shunt coupling capacitance value. The rest of the shunt resistors (R_1 = 5.54 ohm, R_2 = 1e-005 ohm, R_5 = 5.54 ohm and R_6 = 1e-005 ohm) represents the losses effect due to the holes at higher frequencies. In off-state position of the switch (as shown in **Figure 2b**), C_4 = 0.007 pF represents a series switch capacitance value.

After solving equivalent circuit, at a given solution frequency, the impedance in ON position is equal to $Z_{eq} = R_{sw} + iwL_s = 3.98 + 2.52i$ (in ohm) and in OFF position, the capacitance value calculated is C_c = 2.74 femto-Farad. These values are useful to define the insertion loss and isolation. Except the characteristics impedance, other circuit values are permitted to vary the isolation and insertion loss results calculations [48]. Here the actual equivalent circuit of proposed structure was generated by using HSPICE model file. The HSPICE file from HFSS was exported in

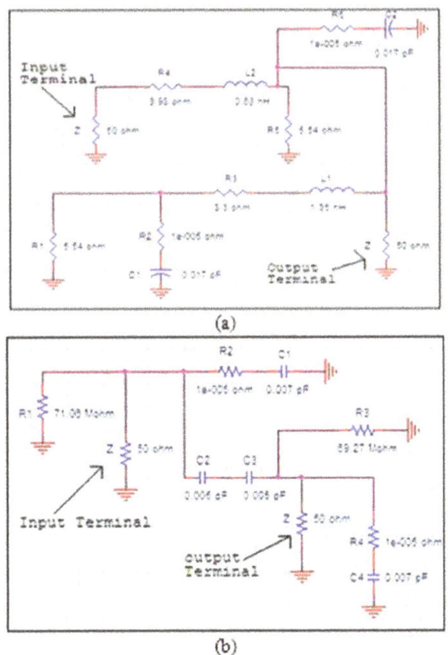

Figure 2. Equivalent R-L-C circuit of metal-contact MEMS switch in (a) on-position (b) off-position.

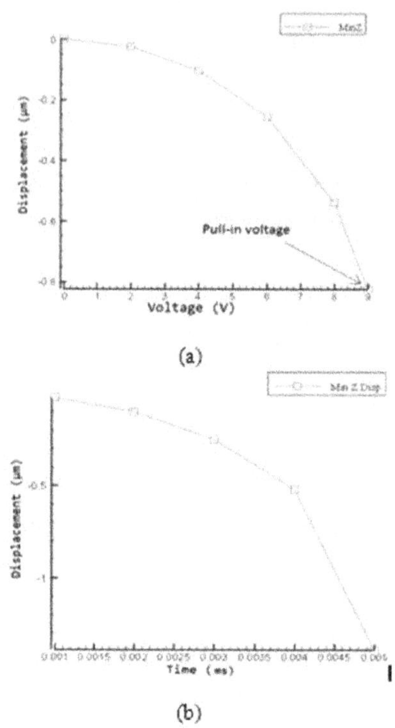

Figure 3. (a) Normalized gap height/displacement (in z-direction) versus applied voltage (pull-in voltage characteristics) and (b) transient response of switch.

Advanced Design System (ADS) software to validate the results through equivalent circuit and layout approach. The generated lumped LCR model of switch in ADS was simulated again by setting 50 ohm impedance at input and output terminal for the verification of post-electromagnetic parameters and further comparison shows similar results as generated by HFSS.

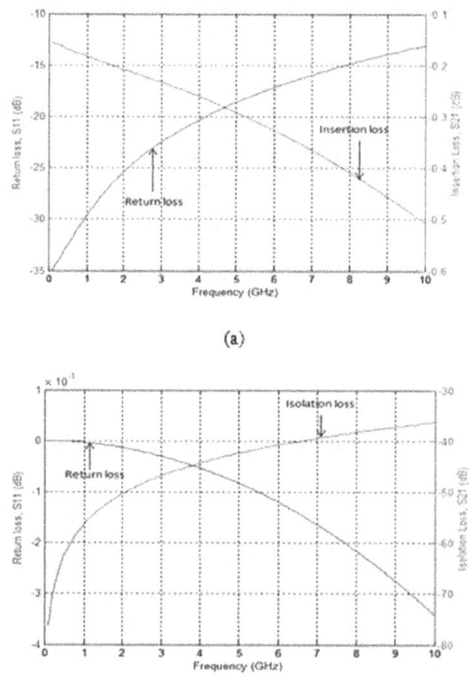

Figure 4. (a) Simulated S-parameter (insertion loss and RL) of switch in on-state and (b) simulated S-parameter (isolation and RL) of switch in off-state (S_{11} and S_{21}).

The mechanical movement depends on spring constant of designed beam structure and lower spring constant is required for better switch operation [49]. At pull-in voltage (V_p = 9.1 V), an electrostatic force is generated between actuation and beam electrode. As a result, beam metal-part makes a contact with transmission line and RF signal passes through this path. The plots of V_p and squeezed-film damping effect in the form of displacement time (t_s) are shown in **Figure 3**. The simulated S-parameters from 0.1 to 10 GHz are shown in **Figure 4**. In ON-position, the insertion and return losses (RL) are 0.15 to 0.51 dB and 34.94 to 13.10 dB, respectively. In OFF-position, the isolation is 75.96 to 35.83 dB.

4. Pattern search optimization method for RF switches

A PS method [33] searches set-points around the existing point, observing for unique where the experimental value of the cost function is lesser than the value at the existing point. The insertion, isolation and RL of presented microswitch in on-off condition are optimized using PS method. These parameters were improved by varying the physical dimensions (width and length) of transmission lines as shown in **Figures 5** and **6**. The substrate thickness and cantilever dimensions are already optimized during the designing process of the RF switch. Here, the problem statement and objective function are by varying structure dimensions of RF microswitch of transmission line to find the maximum transmission from Wave Port 1 to Wave Port 2 (S21 = >1), define the cost function to be –mag (S(WavePort2, WavePort1)) at some specific frequency. The physical dimensions for optimization purpose must be limited so as to avoid the overlarge size as compared with the substrate area, so the restricted limits are set as (20 μm ≤ (L = length) ≤ 45 μm) and (5 μm ≤ (W = width) ≤25 μm), respectively. During optimization analysis, the target is set to identify maximum condition to achieve aforementioned cost function in both on and off position.

Table 4 summarizes the optimized S-parameters of metal-contact switch at 5 GHz. The PS optimization takes six and eight iterations in on and off-positions to optimize the S-parameters with simulation time of 39 and 38 min, respectively. The minimum cost function is achieved in

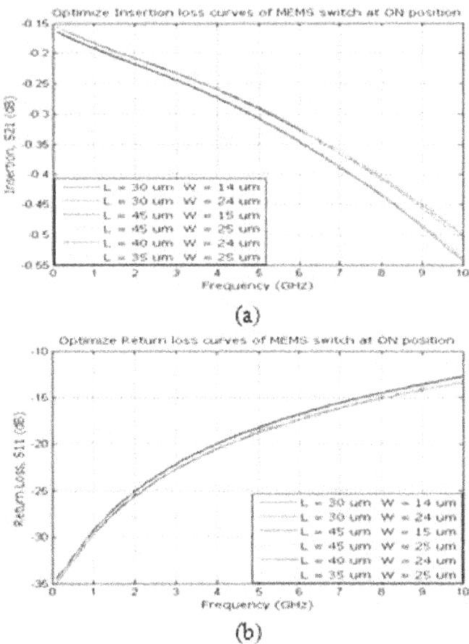

Figure 5. Optimized results at on-position of switch: (a) insertion and (b) return loss.

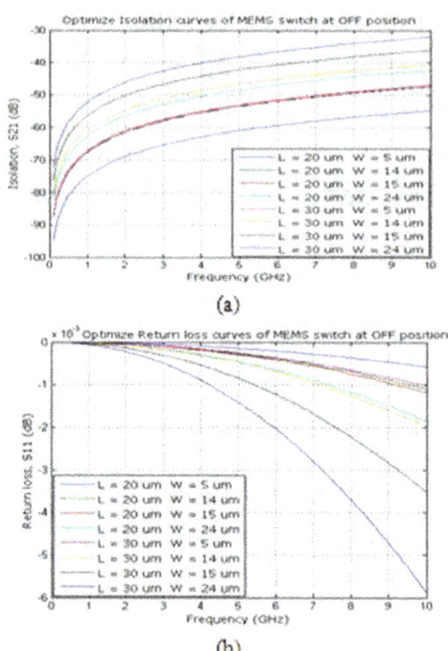

Figure 6. Optimized results at off-position of switch: (a) isolation and (b) return loss.

S-parameters	Without optimization (dB)	With PS optimization (dB)
Insertion loss	−0.32	−0.26
RL in down-state	−17.89	−19.85
Isolation	−41.95	−60.90

Table 4. S-Parameters of MEMS switch at 5 GHz.

fourth iteration in on position and sixth iteration in off position. The optimized value of length and width in on-position is 45 and 25 μm and in off-position of switch is 20 and 5 μm. It has been observed from optimized results that there is major improvement in isolation, reasonable return loss and minor improvement in insertion loss. This is because that PS algorithm follows

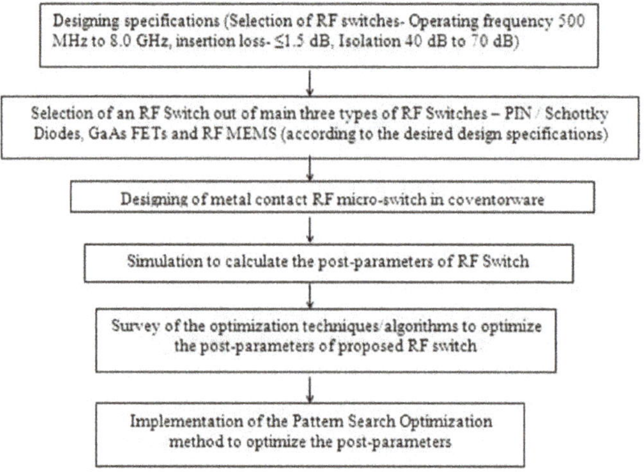

Figure 7. Flow diagram of designing and optimization approach of RF switches.

the designing equations for antenna analysis not includes the transmission line effect and consequently, they are only useful for comparison with simulated data under restricted conditions.

The design approach of RF switches (as discussed in Section 3) with their optimization approach (as discussed in this section) may be shown in the form of flow sequence in **Figure 7**.

5. Defected ground slot reconfigurable antenna

Figure 8 shows the geometry of our proposed defected ground slot antenna which consists of feed-line, patch element and two RF micrometal-contact switches. The antenna was designed on Rogers 4350 substrate of thickness 0.762 mm, relative permittivity $\varepsilon_r = 3.66$ and dielectric loss tangent 0.004. **Table 5** shows all the dimensions of antenna. The positions of switches S_1 and S_2 on antenna were set in a way to operate antenna in dual band. To find the best location of switches on microstrip antenna, the parametric analysis was carried in HFSS software [11, 12].

The proposed RA fabrication and testing set-up broadly contain five different sections which include defected ground planar antenna, RF switches with biasing circuit and voltage regulated power supply to on/off the desired RF switches, vector network analyzer (VNA) and anechoic chamber. The resistive RF microswitch selected for the antenna is same as designed and optimized in this article. For proof of concept, its equivalent RF switches (CSWA2-63DR+) having almost similar isolation and matching of 50 ohm was considered. Although small variation in results are noticed in ON-position of switches at antenna, as compared to simulated results, this is just because of different insertion loss value of tested switch. It consumes very low power (in μW) and typical supply current range is 18 μA. Further, the RF switch used here has operating bandwidth lying from 0.5 to 6.5 GHz. A separate laminate sheet of

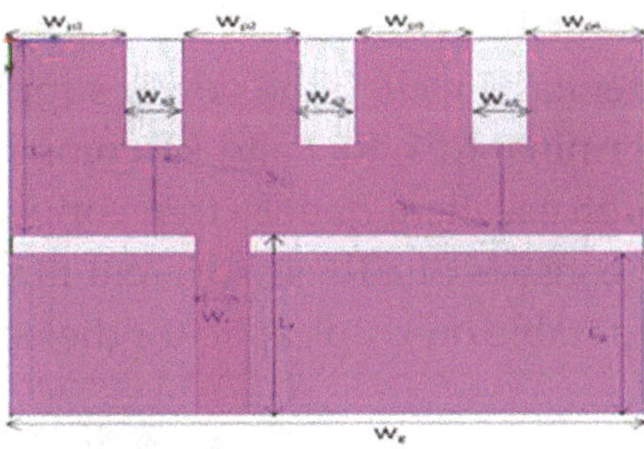

Figure 8. Dual band slot antenna with two metal-contact MEMS switches.

Parameters	Lf	Wf	Lg	Wg	Lp	Wp1
	10	1.8	9	22	11	4
Parameters	**Wp2**	**Wp3**	**Wp4**	**Ws1**	**Ws2**	**Ws3**
	4	4	4	2	2	2

Table 5. Antenna dimensions (in mm).

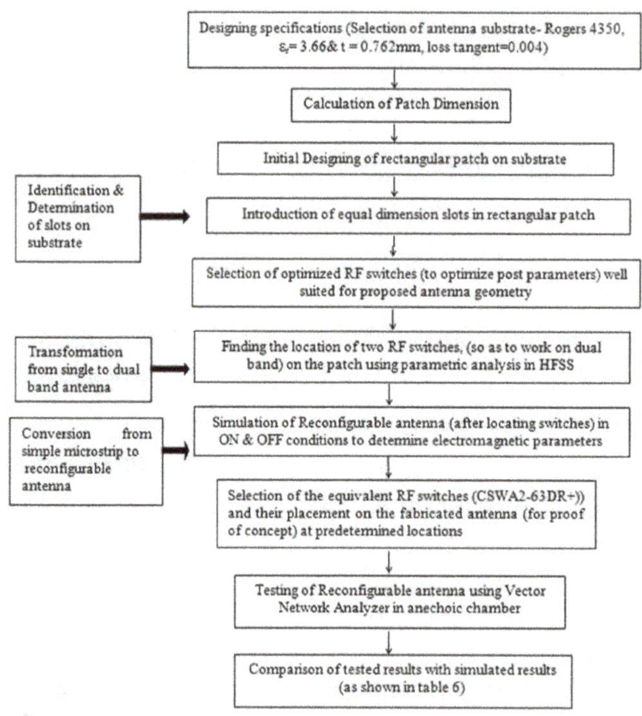

Figure 9. Flow diagram of designing approach of reconfigurable antenna.

Roger RO4350 with same substrate thickness (0.762 mm) as that of defected ground antenna was used for integration of RF switches.

The design approach of reconfigurable antenna (as discussed in this section) may be shown in the form of flow sequence in **Figure 9**.

Figure 10 shows the laminate sheet that consists of 10 RF switches and only two switches (*S*1-*S*2) were used in this work. This PCB carries the DC bias lines and pads that were designed for further connection with antenna as well to the power supply. The gold-plated SMA connector of 50 ohm was used at terminal RF1, RF2 and RF common which further connected to defected ground antenna through excellent quality thin copper wires having diameter 0.025 mm. Each RF switching circuit consists of a RF switch and three equal valued capacitors (47 pF) for blocking DC. The highly stable on-chip capacitor dielectrics C0G (NP0) ceramics was

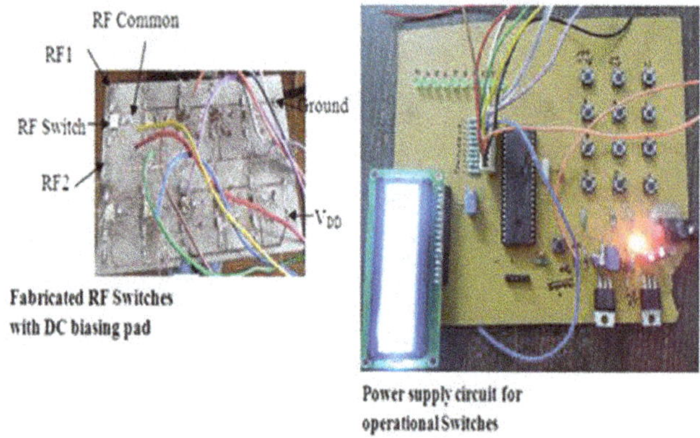

Figure 10. Prototype of RF switches PCB and regulated power supply circuit.

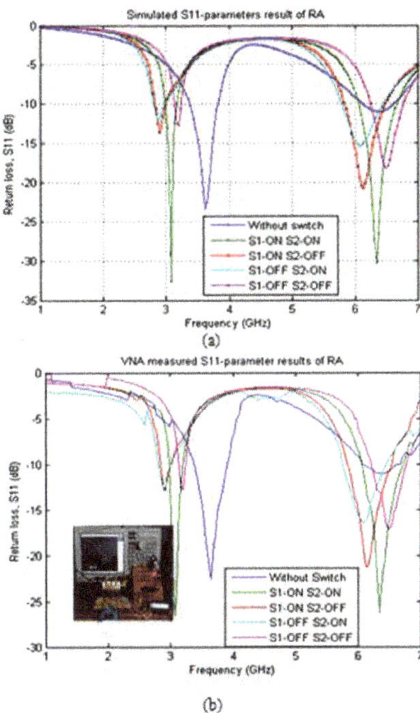

Figure 11. Comparison of simulated and measured RL of proposed RA.

used for biasing circuit. A variable DC regulated power supply producing 1.3–4.92 V was designed to activate RF switches as shown in **Figure 10**.

The electronics switches were integrated on power supply PCB which activates the desired RF switches through 40 pin Atmel microcontroller. The measurements of RA electromagnetic characteristics were done with the help of Agilent Technologies E5071C in an anechoic chamber. The two RF switches provide four possible switching states, that is, on-on, on-off, off-on and off-off. The testing result shows that in on-on state, antenna demonstrates best resonance performance as shown in **Figure 11** and **Table 6**.

Switch-state	Resonating freq. (GHz)		RL (S_{11}) in dB		Bandwidth (MHz)	
	Simulated	Measured	Simulated	Measured	Simulated	Measured
Antenna without switch	3.62	3.58	−23.29	−22.50	300	294
S_1-ON	3.08	3.06	−32.54	−27.20	121	118
S_2-ON	6.34	6.40	−30.25	−26.32	392	384
S_1-ON	2.89	2.85	−13.58	−12.92	151	148
S_2-OFF	6.13	6.08	−20.75	−21.10	392	386
S_1-OFF	2.87	2.86	−11.98	−12.55	181	178
S_2-ON	6.09	6.14	−15.37	−15.18	573	569
S_1-OFF	3.17	3.20	−12.50	−13.10	90	88
S_2-OFF	6.49	6.50	−18.12	−16.78	271	266

Table 6. Comparison of S-parameter results of antenna with and without RF switches.

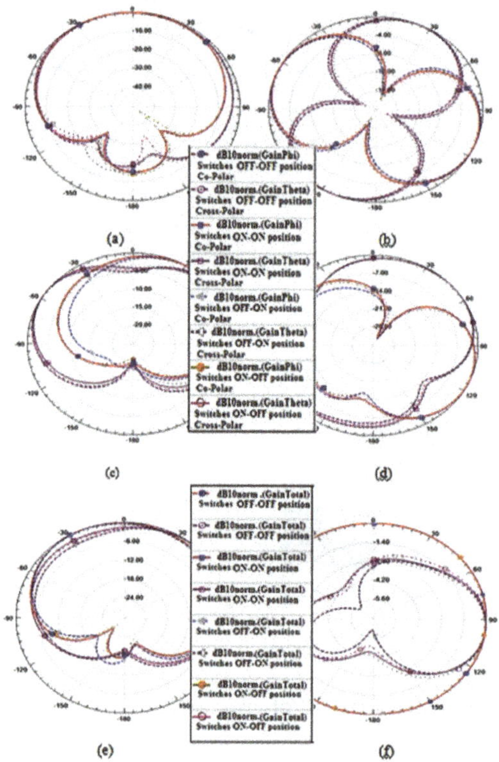

Figure 12. Measured normalized co-polar, cross-polar and total gain radiation patterns in E-plane and H-plane for the reconfigurable antenna at different switching configuration (a) XZ-plane (phi = 0), (b) XY-plane (theta = 0), (c) YZ-plane(phi = 90), (d) gainPhi and gainTheta (theta = 90), (e) gain total (phi = 0, theta = 90) and (f) gain total (theta = 0, theta = 90).

Radiation pattern characteristics—The measured radiation pattern performances for the RA at different switching configuration plotted above are included next. Measured normalized relative power patterns, that is, YZ-, XZ- and XY- cut patterns are shown in **Figure 12a–d** at solution frequency 4 GHz are obtained. The combined magnitude of the electric field components in the desired polarization is shown next in **Figure 12e–f**. A figure-eight pattern in XY-plane (as shown in **Figure 12b**) signifies a dipole type of radiation pattern. Reconfigurable antenna as expected achieved fairly normalized omni-directional patterns. From 3D radiation pattern plot, it has been observed that the antenna behaves directional in the elevation plane and non-directional in the azimuth plane. Further, considering all four possible configurations (on and off) of the switches on antenna, the RA was showing well-behaved linearly polarized characteristic as axial ratio (E_y/E_x) is above one at theta equal to zero.

6. Conclusions

In this chapter, initially a metal-contact microswitch for application of reconfigurable defected ground antenna has been presented, simulated and optimized. Further, the circuit model of RF switch has also been used to extract the capacitance, resistance and inductance parameters by using ADS to validate insertion loss and isolation results. For optimization of S-parameters, pattern search (PS) optimization algorithm has been used after reviewing many of the commonly used optimization techniques. After PS optimization, the RF switch shows significant improvements in insertion loss and isolation at 0.1–10 GHz. Further, the proposed metal-contact microswitch when introduced on defected ground slot antenna and fabricated antenna showed dual-band characteristics as well as reduction in size. From wireless industrial application point of view, the proposed compact reconfigurable antenna with RF switches aiming towards the future wireless miniature devices is suitable for IEEE S- and C-bands. In other words, the proposed reconfigurable antenna finds suitable applications in vehicular and wireless technologies.

Acknowledgements

This work was supported by National Program on Micro and Smart Systems (NPMASS) and also MANCEF, New Mexico, USA, along with coventor organization for providing coventorware, comsol and other useful softwares.

Author details

Paras Chawla[1]* and Rohit Anand[2]

*Address all correspondence to: dr.paraschawla@cgc.edu.in

1 Electronics and Communication Engineering Department, Chandigarh Engineering College Landran, Greater Mohali, Punjab, India

2 Electronics & Communication Engineering Department, G. B. Pant Government Engineering College, New Delhi, India

References

[1] Stefanini R, Chatras M, Blondy P, Rebeiz GM. Miniature MEMS switches for RF applications. Journal of Microelectromechanical Systems. 2011; 20(6): 1324–1335. doi:10.1109/jmems.2011.2170822

[2] Park JH. Fabrication and measurements of direct contact type RF MEMS switch. IEICE Electronics Express. 2012; 4(10): 319–325. doi:10.1587/elex.4.319

[3] Patel CD, Rebeiz GM. A high-reliability high-linearity high-power RF MEMS metal-contact switch for DC–40-GHz applications. IEEE Transactions on Microwave Theory and Techniques. 2012; 60(10): 3096–3112. doi:10.1109/tmtt.2012.2211888

[4] Lucyszyn S, editor. Advanced RF MEMS. Cambridge: Cambridge University Press; 2010. 440 p. doi:10.1017/CBO9780511781995

[5] Lahiri SK, Saha H, Kundu A. RF MEMS switch: An overview at a glance. In: Proceedings of 4th International Conference on Computers and Devices for Communication; 14–16 December 2009; Kolkata. India: IEEE, 2009. p.1–5.

[6] Rebeiz GM, Muldavin JB. RF MEMS switches and switch circuits. IEEE Microwave Magazine. 2001; 2(4): 59–71. doi:10.1109/6668.969936

[7] Muldavin JB. Design and analysis of series and shunt MEMS switches [thesis]. The University of Michigan; 2001.

[8] Oberhammer J. Novel RF MEMS switch and packaging concepts [thesis]. Royal Institute of Technology. Sweden; 2004.

[9] Rebeiz GM. RF MEMS: Theory, Design and Technology. Wiley; New Jersey, USA 2003. 512 p.

[10] Razavi A, Forooraghi K. Thinned arrays using pattern search algorithms. Progress in Electromagnetics Research. 2008; 78: 61–71. doi:10.2528/pier07081501

[11] Chawla P, Khanna R. A novel design and optimization approach of RF MEMS switch for reconfigurable antenna using ANN method. In: Proceedings of International Conference on Communications, Devices and Intelligent Systems; 28–29 December 2012; Kolkata. India: IEEE, 2012. pp. 188–191.

[12] Chawla P, Khanna R. Optimization algorithm of neural network on RF MEMS switch for wireless and mobile reconfigurable antenna applications. In: Proceedings of Second IEEE International Conference on PDGC; 6–8 Dec 2012; Solan. India: IEEE, 2012. pp. 735–740.

[13] Anagnostou DE, Zheng G, Chryssomallis MT, Lyke JC, Ponchak GE, Papapolymerou J, Christodoulou CG. Design, fabrication and measurements of an RF MEMS based self-similar reconfigurable antenna. IEEE Transactions on Antennas and Propagation. 2006; 54(2): 422–432. doi:10.1109/tap.2005.863399

[14] Li Y, Li W, Ye Q. Compact reconfigurable UWB antenna integrated with SIRs and switches for multimode wireless communications. IEICE Electronics Express. 2012; 9(7): doi:10.1587/elex.9.629

[15] Chen YB, Chen TB, Jiao YC, Zhang FS. A reconfigurable microstrip antenna with switchable polarization. Journal of Electromagnetic Waves and Applications. 2006; 20(10): 1391–1398. doi:10.1163/156939306779276820

[16] Majid HA, Rahim MKA, Hamid MR, Ismail MF. Frequency and pattern reconfigurable yagi antenna. Journal of Electromagnetic Waves and Applications. 2012; 26: 379–389. doi:10.1163/156939312800030893

[17] Ding X, Wang B. A millimeter-wave pattern-reconfigurable antenna with a reconfigurable feeding network. Journal of Electromagnetic Waves and Applications. 2013; 27(5): 649–658. doi:10.1080/09205071.2013.759520

[18] Yeung SH, Man KF. Multiobjective optimization. IEEE Microwave Magazine. 2011; 12(6): 120–133. doi:10.1109/mmm.2011.942013

[19] Yeung SH, Chan WS, Ng KT, Man KF. Computational optimization algorithms for antennas and RF/Microwave circuit designs: An overview. IEEE Transactions on Industrial Informatics. 2012; 8(2): 216–227. doi:10.1109/tii.2012.2186821

[20] Dorigo M, Birattari M, Stutzle T. Ant colony optimization. IEEE Computational Intelligence Magazine. 2006; 1(4): 28–39. doi:10.1109/mci.2006.329691

[21] Hashmi A, Gupta D, Goel N, Goel S. Comparative study of bio-inspired algorithms for unconstrained optimization problems. In: Proceedings of Second International Conference on Advances in Electronics, Electrical & Computer Engineering; 2013; Dehradun. India: 2013. pp. 138–142.

[22] Pal SK, Rai CS, Singh AP. Comparative study of firefly algorithm and particle swarm optimization for noisy non-linear optimization problems. International Journal of Intelligent Systems and Applications. 2012; 4(10): 50–57. doi:10.5815/ijisa.2012.10.06

[23] Rajabioun R. Cuckoo optimization algorithm. Applied Soft Computing. 2011; 11(8): 5508–5518. doi:10.1016/j.asoc.2011.05.008

[24] Yang XS, Gandomi AH. Bat algorithm: A novel approach for global engineering optimization. Engineering Computations. 2012; 29(5): 464–483. doi:10.1108/02644401211235834

[25] Goudos SK, Siakavara K, Sahalos JN. Novel spiral antenna design using artificial bee colony optimization for UHF RFID applications. IEEE Antennas and Wireless Propagation Letters. 2014; 13: 528–531. doi:10.1109/lawp.2014.2311653

[26] Hosseini HS. Otsu's criterion-based multilevel thresholding by a nature-inspired metaheuristic called galaxy-based search algorithm. In: Third World Congress on Nature and Biologically Inspired Computing; 19–21 October 2011; Salamanca: 2011. pp. 383–388.

[27] Yang XS. Harmony search as a metaheuristic algorithm. In: Geem ZW, editor. Music-Inspired Harmony Search Algorithm. Springer-Berlin Heidelberg; 2009. p. 1–14. doi:10.1007/978-3-642-00185-7_1 (Studies in Computational Intelligence, Springer Berlin, Vol. 191, pp.1-18, 2009).

[28] Guney K, Onay M. Optimal synthesis of linear antenna arrays using a harmony search algorithm. Expert Systems with Applications, 2011; 38(12): 15455–15462. doi:10.1016/j.eswa.2011.06.015

[29] Singh U, Kamal TS. Synthesis of thinned planar concentric circular antenna arrays using biogeography-based optimisation, IET Microwaves, Antennas & Propagation. 2012; 6(7): 822–829. doi:10.1049/iet-map.2011.0484

[30] Storn R, Price K. Differential evolution—a simple and efficient heuristic for global optimization over continuous spaces. Journal of Global Optimization. 1997; 11(4): 341–359. doi:10.1023/A:1008202821328

[31] Kirkpatrick S, Gelatt Jr. CD, Vecchi MP. Optimization by simulated annealing. Science, New Series, 1983; 220(4598): 671–680.

[32] Karimkashi S, Kishk AA. Invasive weed optimization and its features in electromagnetic. IEEE Transactions on Antennas and Propagation. 2010; 58(4): 1269–1278. doi:10.1109/tap.2010.2041163

[33] Gunes F, Tokan F. Pattern search optimization with application on synthesis of linear antenna arrays. Expert Systems with Applications. 2010; 37(6): 4698–4705. doi:10.1016/j.eswa.2009.11.012

[34] Filho VAD. Portfolio Management Using Value at Risk: A comparison between genetic algorithms and particle swarm optimization [thesis]. Erasmus Universiteit Rotterdam; 2006.

[35] Teodorovic D. Bee Colony Optimization. Innovations in Swarm Intelligence. Springer Berlin Heidelberg, 2009; 248: 39–60. doi:10.1007/978-3-642-04225-6_3

[36] Singh U, Kamal TS. Design of non-uniform circular antenna arrays using biogeography-based optimization. IET Microwaves, Antennas & Propagation. 2011; 5(11): 1365–1370. doi:10.1049/iet-map.2010.0204

[37] Price KV, Storn RM, Lampinen JA. Differential Evolution: A Practical Approach to Global Optimization. Springer-Verlag Berlin Heidelberg; 2005. 539 p. doi:10.1007/3-540-31306-0

[38] Kumbharana SN, Pandey GM. A comparative study of ACO, GA and SA for solving travelling salesman problem. International Journal of Societal Applications of Computer Science. 2013; 2(2): 224–228.

[39] Peng S, Ouyang AJ, Zhang JJ. An adaptive invasive weed optimization algorithm. International Journal of Pattern Recognition and Artificial Intelligence. 2015; 29(2): 1559004 (1–19). doi:10.1142/S0218001415590041

[40] Samsami R. Comparison between genetic algorithm (GA), particle swarm optimization (PSO) and ant colony optimization (ACO) techniques for NO_X emission forecasting in Iran. World Applied Sciences Journal. 2013; 28(12): 1996–2002. doi:10.5829/idosi.wasj.2013.28.12.1155

[41] Valian E, Mohanna S, Tavakoli S. Improved cuckoo search algorithm for feedforward neural network training. International Journal of Artificial Intelligence & Applications. 2011; 2(3): 36–43. doi:10.5121/ijaia.2011.2304

[42] Duan Y. A hybrid optimization algorithm based on bat and cuckoo search. Advanced Materials Research. 2014; 926–930: 2889–2992. doi:10.4028/www.scientific.net/AMR.926-930.2889

[43] Wang X, Gao XZ, Zenger K. An Introduction to Harmony Search Optimization Method. Springer; 2015. 88 p. Illus. DOI 10.1007/978-3-319-08356-8_1

[44] Srikanth D, Barai SV. Structural optimization using harmony search algorithm. Soft Computing in Industrial Applications. Springer-Verlag Berlin Heidelberg. 2010; 75:61–69. doi:10.1007/978-3-642-11282-9_7

[45] Ammu PK, Sivakumar KC, Rejimoan R. Biogeography-based optimization. International Journal of Electronics and Computer Science Engineering. 2012; 2(1): 154–160.

[46] Liu AQ, Yu AB, Karim MF, Tang M. RF MEMS switches and integrated switching circuits. Journal of Semiconductor Technology and Science. 2007; 7(3): 166–176.

[47] Liu AQ. RF MEMS switches and integrated switching circuits. 1st Ed. U.S.: Springer; 2010. 264 p. doi:10.1007/978-0-387-46262-2

[48] Singh K, Nagachenchaiah K, Bhatnagar D. Electromagnetic modelling of conductor-backed CPW based RF MEMS capacitive shunt switch. International Journal of Electronics. 2009; 96(8): 887–893. doi:10.1080/00207210902851498

[49] Garg A, Chawla P, Khanna R. A Novel Approach of RF MEMS Resistive Series Switch for Reconfigurable Antenna. In: Proceedings of Annual IEEE Int. Conf.on Emerging Research Areas and 2013 International Conference on Microelectronics, Communications and Renewable Energy; 4–6 June 2013; Kanjirapally: 2013. pp. 1–6

7

Recent Computer-Aided Design Techniques for Rectangular Microstrip Antenna

Sudipta Chattopadhyay and

Subhradeep Chakraborty

Abstract

In modern microwave systems, rectangular microstrip patch antennas (RMPAs) are probably the most investigated topics among the planar antennas. There are several methods available in literature, for designing and analyzing such antennas, but most of them are very complex and give only approximate results. In this chapter, we have discussed the most accurate and updated computer-aided design (CAD) formulations related to probe-fed RMPA for computing its fundamental input characteristics (reso-nant frequency and input impedance) and improving radiation characteristics, i.e. gain and polarization purity (the parameter that signifies how much an RMPA is free from spurious modes). These formulations have evolved in the last decades and have been validated against numerous simulations and measurements. The present CAD formulas for resonant frequency and input impedance can accurately address a wide range of RMPA with patch width to patch length ratio (*W*/*L*) from 0.5 to 2.0, a substrate having thickness up to 0.23 λ_g where λ_g is the guide wavelength and relative permittivity (ε_r) ranging over 2.2–10.8. The role of a finite air gap on resonant frequency and gain of an RMPA have also been presented. The chapter will be surely useful to antenna designers to achieve a concrete understanding of the RMPA theory.

Keywords: rectangular microstrip antenna, resonant frequency, input impedance, gain, polarization purity, grounded microstrip patch

1. Introduction

'Microstrip antennas', the class of antennas which has been capturing the attention of the antenna research community for the last 63 years, starting from the 3rd Symposium on the US Air Force Antenna Research and Development Program, was proposed firstly by Deschamp and Sichak [1] in 1953. In [1], they proposed a microstrip feeding network for a waveguide system which comprised of 300 waveguide horn antennas. But, the 'microstrip patch', to which all the researchers associated to the antenna theory are familiar with, was theoretically analyzed by Howell in [2] and applied into practical applications by Munson in [3] for the first time. Still, the credit of the authors of the work [1] was to foster a very much new discipline of antenna engineering, and in addition to this, they highlighted the related performance superiority of this new antenna over the other commercially available conventional antenna. From 1972 onwards, researchers started to understand a microstrip patch in two approaches: (i) treating the patch as a lossy and open resonator cavity [2] and (ii) as an extended section of a microstrip line [3].

In the last four decades, several books [4–11] and collection of research papers [12, 13] have been published unfolding rigorous analytical and numerical techniques dealing with microstrip antennas. The computational methods like FDTD, FEM and MOM are very much versatile in nature particularly in analyzing irregular-shaped patch geometries with huge various types of substrates, but any of them give neither any physical insights into the radiation mechanism of the antenna nor any closed-form design formulations which are utmost necessary to any practicing antenna engineer, researcher, academician or scientist. In this context, cavity resonator model appears to be more effective than the other available methods to estimate the fundamental input characteristics (i.e. resonant frequency and input impedance) and to improve the crucial radiation characteristics (i.e. gain and polarization purity) for commonly available microstrip antennas of regular geometries with thin substrate. This method not only improvises the design steps of antenna design but also aids in achieving better performance.

In this chapter, the authors have presented comprehensive electromagnetic analyses on the fundamental input characteristics (i.e. resonant frequency and input impedance) and radiation characteristics (i.e. gain and polarization purity) of rectangular microstrip antennas with conventional and suspended geometries in light of the versatile cavity model method and discussed some improved and closed-form computer-aided design (CAD) formulas. Unlike other theories and work, the present CAD formulas can accurately address a wide range of aspect ratio or patch width to patch length ratio (W/L) from 0.5 to 2.0, a substrate having thickness up to 0.23 λ_g where λ_g is the guide wavelength and relative permittivity (ε_r) ranging over 2.2–10.8. (Discuss Novelty) A coaxially fed RMPA with length L and width W on a substrate (ε_r) above a variable air-gap over the ground plane is shown in **Figure 1**.

Air gap over the ground plane is shown in **Figure 1**. The variable air-gap height h_1 can be decreased to zero to achieve the conventional form. The fringing of the electric fields at the radiating and non-radiating edges of the rectangular patch is taken into account in terms of ΔL and ΔW, respectively. The present CAD formulas were introduced firstly in [14] by Chattopadhyay et al.

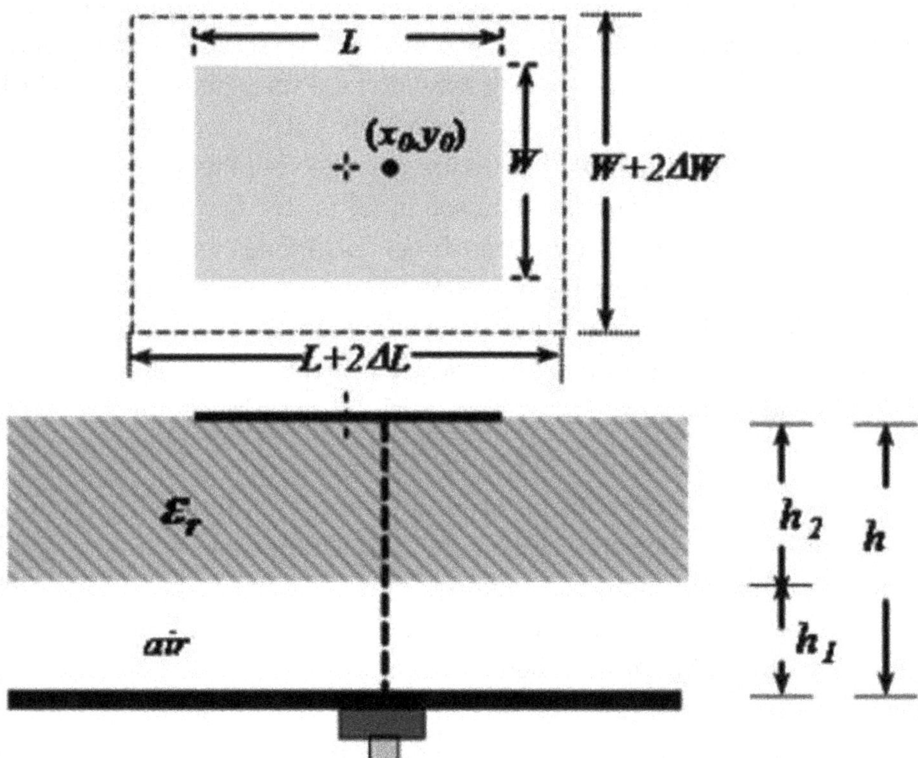

Figure 1. Schematic diagram of a coax-fed rectangular microstrip patch antenna (RMPA).

2. Input characteristics

2.1. Resonant frequency

After the work of Howell [2], Hammerstad [15] proposed comprehensive CAD formulations on resonance characteristics of an RMPA using the cavity resonator model (CRM) method. Till now, several theoretical analyses employing CRM method [6, 16–18], transmission-line method [19, 20], method of moments [21, 22] and integral equation technique [23] are available in literature. But the CAD formulas presented in [6, 15] are found to be the most popular for the design purpose.

Nevertheless, a close inspections into these works show that the formulas available in [6, 15] can provide a reasonably good approximation only when the patches have aspect ratio (W/L) near to 1.5 and the substrate thickness is lower than 0.02 λ_g, where λ_g is the guide wavelength at the resonant frequency f_r. Also, the computed f_r values using formulas in [6, 15] show errors in comparison with the measurements which can be found in [24, 25]. Those works [24, 25] used around 0.04 λ_g to 0.23 λ_g substrate thickness. The use of an air gap in between the substrate and the ground plane has been found in [26–34] which helps in achieving tunability of an RMPA and enhancing its impedance bandwidth. Earlier, a cavity model was also discussed in detail in [5, 11]. In this section, authors have emphasized on a better CAD formulation as found in an earlier work of one of the present authors in [14] using the well-known CRM method called quasi-static approach [11] to estimate more accurately the dominant and the higher-order resonances in an RMPA with and without air gap. Here, ΔL and ΔW

have been considered as a function of the aspect ratio. In [14] Chattopadhyay et al. considered an equivalent circular patch with radius a, effective radius a_{eff} and same resonant frequency as that of the RMPA. This helps to establish a relationship among the fringing parameters ΔL, ΔW and a_{eff} by equating the zero-order resonant frequencies of both patches as found in one of the earlier works of Chattopadhyay in [14, 35].

From [35–38] we can write

$$f_{0,r} = \frac{c}{2L\sqrt{\varepsilon_r}} \tag{2.1}$$

where c is the velocity of light in free space. Following the earlier work of Chattopadhyay in [14, 35], we can assume that both antennas of same resonant frequencies have same circumference. Therefore, we can write

$$(\mathrm{W} + \mathrm{L}) = \pi \mathrm{a} \tag{2.2}$$

$$(\mathrm{L} + 2\Delta\mathrm{L}) + (\mathrm{W} + 2\Delta\mathrm{W}) = \pi a_{eff} \tag{2.3}$$

In Eq. (2.3) α is the first zero of the derivative of the Bessel function of order $n = 1$. Now, a_{eff} is the effective radius of a circular patch due to fringing electric fields as found in [27]:

$$a_{eff} = a\sqrt{1+q} \tag{2.4}$$

where q is the fringing factor calculated from [27] as

$$q = u + v + uv \tag{2.5}$$

and

$$u = \left(1 + \varepsilon_{re}^{-1}\right)\frac{4}{\pi a/h} \tag{2.6}$$

$$v = \frac{2}{3t} \times \frac{\ln(p)}{(8 + {\pi a}/{h})} + \frac{({1}/{t} - 1)}{(4 + 2.6{a}/{h} + 2.9{h}/{a})} \tag{2.7}$$

$$t = 0.37 + 0.63\varepsilon_{re} \tag{2.8}$$

$$p = \frac{1 + 0.8\left(\frac{a}{h}\right)^2 + \left(\frac{0.31a}{h}\right)^4}{1 + 0.9\frac{a}{h}} \tag{2.9}$$

u, v, t and p all are dummy variables.

$$\varepsilon_{re} = \frac{\varepsilon_r\left(1 + \frac{h_1}{h_2}\right)}{(1 + \varepsilon_r h_1/h_2)} \tag{2.10}$$

where ε_{re} is the equivalent permittivity of the two-layer dielectric medium (**Figure 1**) having a total thickness $h = (h_1+h_2)$.

Solving (2.1)–(2.4) and from one of the previous works of Chattopadhyay in [14, 35], one can write the following relations:

$$L = 1.7a \tag{2.11}$$

$$W = 1.44a \tag{2.12}$$

$$\Delta L + \Delta W = \frac{\pi a\sqrt{1+q} - 1}{2} \tag{2.13}$$

An empirical relation is used to determine ΔW in terms of ΔL for a wide range of W/L values, $2 > W/L > 0.5$ as

$$\Delta W = \Delta L\left(1.5 - \frac{W}{2L}\right) \tag{2.14}$$

and ΔL can be written as given in one of the previous works of Chattopadhyay in [14, 35]

$$\Delta L = \frac{\pi a\left[\sqrt{1+q} - 1\right]}{2\left[2.5 - 0.5\left(\frac{W}{L}\right)\right]} \tag{2.15}$$

Now, using the above Eqs. (2.1)–(2.15) discussed above and as in the previous work of Chattopadhyay in [14, 35], the resonant frequency of an RMPA with a variable air-gap h_1 is found as

$$f_{r,nm} = \frac{c}{2\sqrt{\varepsilon_{r,eff}}}\left[\left(\frac{n}{L+2\Delta L}\right)^2 + \left(\frac{m}{W+2\Delta W}\right)^2\right]^{1/2} \tag{2.16}$$

where $\varepsilon_{r,eff}$ is the effective relative permittivity of the medium below the patch [14, 27, 35].

$$\varepsilon_{\mathrm{r,eff}} = \frac{4\,\varepsilon_{\mathrm{re}}\,\varepsilon_{\mathrm{r,dyn}}}{\left(\sqrt{\varepsilon_{\mathrm{re}}} + \sqrt{\varepsilon_{\mathrm{r,dyn}}}\right)^2} \tag{2.17}$$

From Eq. (2.16) we can found the dominant mode of an RMPA is TM_{10}. ε_{re} is calculated using (2.10), and $\varepsilon_{r,dyn}$ is the dynamic dielectric constant as defined in [17, 28] and can be written as

$$\varepsilon_{r,dyn} = \frac{C_{dyn}\left(\varepsilon = \varepsilon_0\varepsilon_{re}\right)}{C_{dyn}\left(\varepsilon = \varepsilon_0\right)} \tag{2.18}$$

$$C_{dyn} = C_{0,dyn} + C_{e,dyn} \tag{2.19}$$

where C_{dyn} is the total dynamic capacitance of the RMPA and suffixes *0* and *e* denote the main and fringing components, respectively. $C_{0,dyn}$ and $C_{e,dyn}$ are determined as discussed in one of the previous works of the present authors in [14, 35]:

$$C_{0,dyn} = \gamma_n C_{0,stat} \tag{2.20}$$

$$C_{e,dyn} = \frac{1}{\delta} C_{e,stat} \tag{2.21}$$

The values γ_n and δ are as follows:

$$\gamma_n = 1.0 \text{ for } n = 0 \tag{2.22}$$

$$= 0.3525 \text{ for } n = 1$$

$$= 0.2865 \text{ for } n = 2$$

$$= 0.2450 \text{ for } n = 3$$

$$\delta = 1.0 \text{ for } n = 0 \tag{2.23}$$

$$= 2.0 \text{ for } n \neq 0$$

where $C_{0,stat}$ and $C_{e,stat}$ are the static main and static fringing capacitances of the disc, and these are [35]

$$C_{0,stat} = \varepsilon_0 \varepsilon_{re} \frac{\pi a^2}{h} \tag{2.24}$$

$$C_{e,stat} = C_{0,stat} q \tag{2.25}$$

Figure 2 shows the computed as obtained from the previous work of Chattopadhyay in [14, 35] and measured [24] resonant frequencies of an RMPA as a function of substrate thickness h_2. The three measured values for thin and thick substrates show close agreement with proposed formulations.

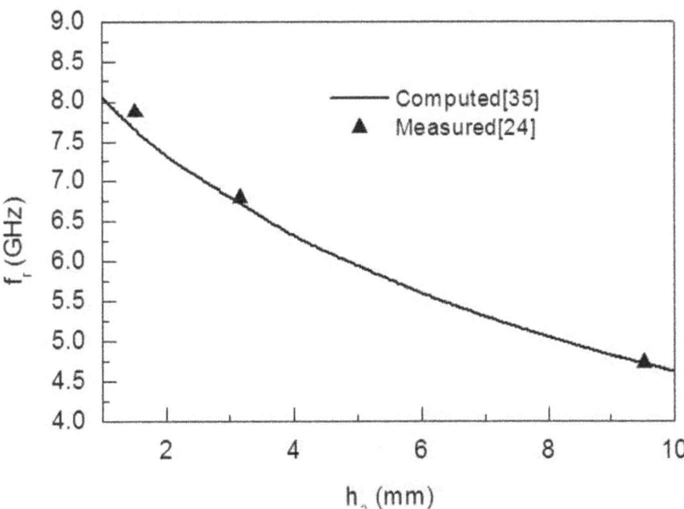

Figure 2. Resonant frequency of the dominant mode as function of substrate thickness of the RMPA, $\varepsilon_r = 2.33$, $L = 11$ mm, $W = 17$ mm, $W/L = 1.54$ [35].

In [14], Chattopadhyay et al. have also shown the close agreement between the values obtained using MOM and their theory.

Figures 3 and **4** show the computed (from the previous work of Chattopadhyay in [14, 35]), simulated and measured resonant frequencies of an RMPA with $W/L = 1.5$ having variable air-gap heights.

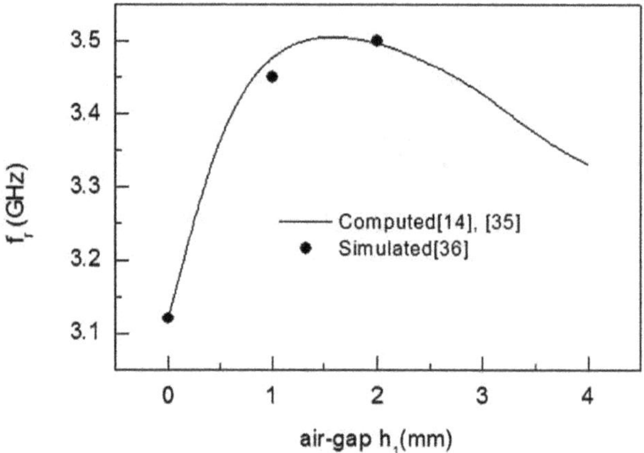

Figure 3. Resonant frequency of RMPA versus air-gap height. $\varepsilon_r = 2.2$, L = 30 mm, W = 45 mm, W/L = 1.5, substrate thickness $h_2 = 1.575$ mm [35].

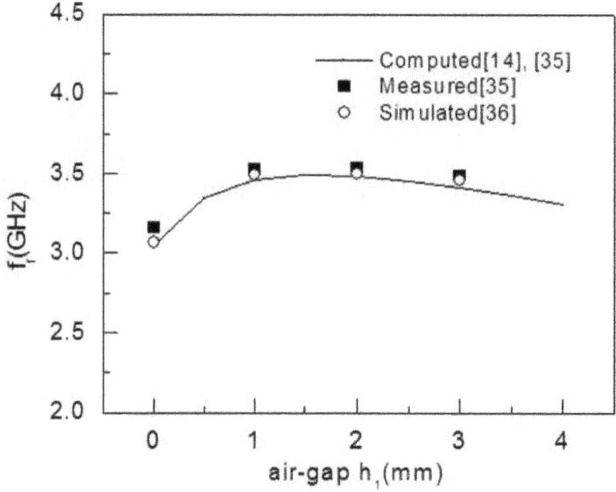

Figure 4. Dominant mode resonant frequency of RMPA versus air-gap height. $\varepsilon_r = 2.33$, L = 30 mm, W = 45 mm, $h_2 = 1.575$ mm [35].

The simulated and measured values ranging from no air gap to an air gap of 4 mm show good agreement with the present formulations. The tunability of the RMPA as a function of the air gap height has also been studied. **Table 1** compares the resonant frequencies as computed using the presented formulations (as given in the previous work of Chattopadhyay in [14, 35]) with that of Hammerstad [15], James et al. [6] and Chew and Liu [23] for different sets of RMPAs with $W/L \approx 1.5$ and for electrical thickness ranging from 0.037 to 0.166 λ_g. The presented formulations show very close agreement with measured data [24] with an average percentage error of (1.39%) for the present formulations.

Length, L (mm)	W/L	Normalized thickness (h_2/λ_d)	Measured fr (GHz) [24]	Computed fr (GHz) [15]	Computed fr (GHz) [6]	Computed fr (GHz) [23]	Computed fr (GHz) (from the earlier work of one of the present authors in [14, 35])
38	1.5	0.037	2.31	2.38 (3.03%)	2.30 (0.4%)	2.37 (2.6%)	2.32 (0.4%)
30.5	1.49	0.047	2.89	2.90 (0.3%)	2.79 (3.4%)	2.90 (0.3%)	2.83 (2%)
19.5	1.51	0.068	4.24	4.34 (2.35%)	4.11 (3.06%)	4.32 (1.88%)	4.18 (1.4%)
13	1.5	0.094	5.84	6.12 (4.79%)	5.70 (2.39%)	6.07 (3.93%)	5.86 (0.3%)
11	1.54	0.110	6.80	7.01 (3.08%)	6.47 (4.85%)	6.90 (1.5%)	6.65 (2.2%)
9	1.55	0.125	7.70	8.19 (6.36%)	7.46 (3.11%)	7.87 (2.2%)	7.73 (0.38%)
8	1.50	0.141	8.27	9.01 (8.94%)	8.13 (1.7%)	8.39 (1.45%)	8.50 (2.7%)
7	1.50	0.148	9.14	9.97 (9.01%)	8.89 (2.73%)	8.69 (4.92%)	9.3 (1.75%)
6	1.50	0.166	10.25	11.18 (9.07%)	9.82 (4.19%)		10.4 (1.46%)
Average error w.r.t measurement [24]				**5.21%**	**2.87%**	**2.34%**	**1.39%**

Parameters: $\varepsilon_r = 2.33$, $h_2 = 3.175$ mm, for $W/L = 1.5$. Note: here, $\lambda_d = \lambda_g = \lambda_0/\sqrt{\varepsilon_r}$ and λ_0 are wavelengths corresponding to measured frequency.

Table 1. Comparison of the measured [24] and computed dominant mode resonant frequency of an RMPA.

Moreover, in [14] Chattopadhyay et al. showed the versatility of these formulations in accurately predicting the higher-order modes of an RMPA for W/L = 1. In [35], Chattopadhyay has predicted the higher-order modes of an RMPA for W/L = 0.7, 1.2, 1.5, 1.7 extending the work in [14]. One can refer to **Table 1** of one of the previous works of Chattopadhyay et al. in [14] for a closer look into the topic. It is seen that the significant higher-order modes of an RMPA are TM_{01}, TM_{02}, TM_{12}, TM_{20}, TM_{30}, TM_{03}, etc. When W/L = 1, TM_{10} and TM_{01} become degenerate modes. The separation between resonant frequency of dominant TM_{10} mode and that of net higher-order mode TM_{02} is from 2 to $1.25f_{r,10}$ for $0.7 < W/L < 2$ as discussed in [35] by Chattopadhyay. The effect of TM_{02} mode on the radiation characteristics of an RMPA is very detrimental [11, 14, 35], and the two newest techniques for mitigating this issue are discussed later in this chapter (see Section 4.2).

2.2. Input impedance

An RMPA can be represented as an equivalent R-L-C parallel resonant circuit in order to find out its input impedance [11]. Near resonance of the dominant mode and its input impedance can be expressed as [39, 40]

$$Z_{in}(f, x_0) = \frac{R_r}{1 + Q_T^2\left(\bar{f} - \bar{f}^{-1}\right)^2} + j\left[X_f - \frac{R_r Q_T\left(\bar{f} - \bar{f}^{-1}\right)}{1 + Q_T^2\left(\bar{f} - \bar{f}^{-1}\right)^2}\right] \tag{2.26}$$

where $\bar{f} = f/f_r$, f_r is the dominant mode resonant frequency and R_r is the input resistance at resonance as [8].

R_r can be expressed as

$$R_r = \frac{4h}{\pi\lambda_0}\mu\eta_0 Q_T\left(\frac{L + 2\Delta L}{W + 2\Delta W}\right)\cos^2\left(\frac{\pi(0.5L - x_0)}{L + 2\Delta L}\right) \tag{2.27}$$

where η_0 is the intrinsic impedance of free space where $\eta_0 = 377\ \Omega$, x_0 is the distance from the centre of the patch and Q_T is the total quality factor.

Q_T can be expressed in terms of the losses due to radiation (Q_r), dielectric (Q_d) and conductor (Q_c) present in the radiating structure as given in an earlier work of one of the present authors in [40]:

$$Q_T = \left[\frac{1}{Q_r} + \frac{1}{Q_d} + \frac{1}{Q_c}\right]^{-1} \tag{2.28}$$

In this context, another parameter $\varepsilon_{r,n}$ is required to calculate Q_r and Q_d as found in [39]:

$$\varepsilon_{r,n} = \frac{\varepsilon_{reff} + 1}{2} \tag{2.29}$$

Now, Q_r, Q_d and Q_c can be expressed as given in [36, 39], and an earlier work of one of the present authors in [40]:

$$Q_r = \frac{\pi}{4G_r Z_r} \tag{2.30}$$

$$Q_d = \frac{\pi(\varepsilon_r - 1)\sqrt{\varepsilon_{r,n}}}{27.3(\varepsilon_{r,n} - 1)\sqrt{2\varepsilon_{r,n} - 1}}\frac{1}{\tan\delta} \tag{2.31}$$

$$Q_c = h\sqrt{\pi f \mu_0 \sigma} \tag{2.32}$$

where

$$G_r = \frac{W^2}{90\lambda_0^2} \text{ for } W \leq 0.35\lambda_0 \tag{2.33}$$

$$= \frac{W}{120\lambda_0} - \frac{1}{60\pi^2} \text{ for } 0.35\lambda_0 \leq W \leq 2\lambda_0$$

$$= \frac{W}{120\lambda_0} \text{ for } 2\lambda_0 < W$$

and

$$Z_r = \frac{120\pi\left[\frac{W}{h} + 1.393 + 0.667\ln\left(\frac{W}{h} + 1.444\right)\right]^{-1}}{\sqrt{\varepsilon_{r,n}}} \tag{2.34}$$

The same approach is also valid for circular patches, and a detailed discussion on the resonant frequency and input impedance of a circular patch can be found in [11]. From [11], one can find that the dominant mode of a circular patch is TM_{11}. The immediate higher-order modes are TM_{21}, TM_{01}, TM_{31}, etc. The formulas are found to be very accurate in case of substrate with thin and moderate height.

Cavity model analysis of the resonant frequency and input impedance for a 60°–60°–60° equilateral triangular patch is found in [5]. The dominant mode of a triangular patch is TM_{10} [5]. The immediate higher-order modes are TM_{11}, TM_{20}, TM_{21}, etc.

3. Examples

(a)Find out the resonant frequency of an RMPA with length (L) 18.2 mm and width (W) 28 mm, etched on a PTFE substrate of height 1.575 mm with dielectric constant 2.33.

(b)Repeat the problem when 1 mm air gap is introduced between substrate and ground plane.

Solution:

(a) The resonant frequency of the patch can be obtained from Eq. (3.1) as

$$f_{r,nm} = \frac{c}{2\sqrt{\varepsilon_{r,eff}}}\left[\left(\frac{n}{L + 2\Delta L}\right)^2 + \left(\frac{m}{W + 2\Delta W}\right)^2\right]^{1/2} \tag{3.1}$$

As the dominant mode is TM_{10}, $n = 1$ and $m = 0$, and therefore the Eq. (3.1) reduces to

$$f_{r,10} = \frac{c}{2\sqrt{\varepsilon_{r,eff}}}\frac{1}{L + 2\Delta L} \tag{3.2}$$

The expression for effective relative permittivity of the medium below the patch is

$$\varepsilon_{r,eff} = \frac{4\varepsilon_{re}\varepsilon_{r,dyn}}{\left(\sqrt{\varepsilon_{re}} + \sqrt{\varepsilon_{r,dyn}}\right)^2} \tag{3.3}$$

and

$$\varepsilon_{re} - \frac{\varepsilon_r\left(1 + \frac{h_1}{h_2}\right)}{(1 + \varepsilon_r h_1/h_2)} \tag{3.4}$$

As air gap height $h_2 = 0$

$$\varepsilon_{re} = \varepsilon_r = 2.33 \tag{3.5}$$

Fringing factor q is

$$q = u + v + uv \tag{3.6}$$

where

$$u = \left(1 + \varepsilon_{re}^{-1}\right) \frac{4}{\frac{\pi a}{h}} = 0.147 \tag{3.7}$$

$$t = 0.37 + 0.63\varepsilon_{re} = 1.83 \tag{3.8}$$

and

$$p = \frac{1 + 0.8\left(\frac{a}{h}\right)^2 + \left(\frac{0.31a}{h}\right)^4}{1 + 0.9\frac{a}{h}} = 27.86 \tag{3.9}$$

Hence,

$$v = \frac{2}{3t} \times \frac{\ln(p)}{(8 + \pi a/h}) + \frac{(1/t - 1)}{(4 + 2.6a/h + 2.9h/a)} \tag{3.10}$$

Therefore,

$$q = u + v + uv = 0.162$$

$\varepsilon_{r.dyn}$ can be calculated using Eqs. (2.18)–(2.25) and (3.6)–(3.10) as

$$\varepsilon_{r.dyn}=1.94$$

and

$\varepsilon_{r,eff}$=2.12.

The fringing length ΔL may be computed as

$$\Delta L = \frac{\pi a\left[\sqrt{1+q} - 1\right]}{2\left[2.5 - 0.5\left(\frac{W}{L}\right)\right]} \tag{3.11}$$

=1.365.

Therefore, the resonant frequency $f_{r,10}$ becomes

$$f_{r,10} = \frac{c}{2\sqrt{\varepsilon_{r,eff}}} \frac{1}{L + 2\Delta L} = 4.916\,\text{GHz}$$

(b) Now, if h_2 = 1 mm,

$$\varepsilon_{re} = \frac{\varepsilon_r\left(1+\frac{h_1}{h_2}\right)}{(1+\varepsilon_r h_1/h_2)} = 1.56$$

$\varepsilon_{r.dyn} = 1.35$

and

$\varepsilon_{r.eff} = 1.43$

and

$$u = \left(1+\varepsilon_{re}^{-1}\right)\frac{4}{\frac{\pi a}{h}} = 0.278$$

$$t = 0.37 + 0.63\varepsilon_{re} = 1.337$$

$$p = \frac{1+0.8\left(\frac{a}{h}\right)^2 + \left(\frac{0.31a}{h}\right)^4}{1+0.9\frac{a}{h}} = 9.83$$

Hence,

$$v = \frac{2}{3t} \times \frac{\ln(p)}{(8+{}^{\pi a}/_{h})} + \frac{({}^{1}/_{t}-1)}{(4+2.6{}^{a}/_{h}+2.9{}^{h}/_{a})} = 0.0253$$

Therefore,

$$q = u + v + uv = 0.310$$

and hence

$$\Delta L = \frac{\pi a\left[\sqrt{1+q}-1\right]}{2\left[2.5-0.5\left(\frac{W}{L}\right)\right]} = 2.52$$

Therefore, the resonant frequency $f_{r.10}$ becomes

$$f_{r,10} = \frac{c}{2\sqrt{\varepsilon_{r,eff}}}\frac{1}{L+2\Delta L} = 5.37\,\text{GHz}$$

(a) Find out the input resonant resistance at the edge of a square patch with length 30 mm and width 30 mm, etched on a PTFE substrate of height 1.575 mm with dielectric constant 2.33.

(b)Find out the optimum feed position.

Solution:

(a) The input impedance of a patch can be expressed as

$$Z_{in}(f, x_0) = \frac{R_r}{1+Q_T^2\left(\bar{f}-\bar{f}^{-1}\right)^2} + j\left[X_f - \frac{R_r Q_T\left(\bar{f}-\bar{f}^{-1}\right)}{1+Q_T^2\left(\bar{f}-\bar{f}^{-1}\right)^2}\right] \quad (3.12)$$

where $\bar{f} = f/f_r$, f_r being the dominant mode resonant frequency and R_r is the input resistance at resonance at the edge of the patch.

Now, f_r can be obtained as done in the earlier example, and it is found to be f_r = 3.13 GHz.

R_r can be expressed as

$$R_r = \frac{4h}{\pi\lambda_0}\mu\eta_0 Q_T\left(\frac{L+2\Delta L}{W+2\Delta W}\right)cos^2\left(\frac{\pi(0.5L-x_0)}{L+2\Delta L}\right) \tag{3.13}$$

where η_0 = 377 Ω and x_0 is the distance from the centre of the patch and Q_T is the total quality factor, expressed in terms of the losses due to radiation (Q_r), dielectric (Q_d) and conductor (Q_c) present in the radiating structure as

$$Q_T = \left[\frac{1}{Q_r} + \frac{1}{Q_d} + \frac{1}{Q_c}\right]^{-1} \tag{3.14}$$

Now,

$$\lambda_0 = {}^{c}/_{f_r} = 95.84\text{mm} \tag{3.15}$$

ΔL can be obtained as done in the earlier example, and it is ΔL = 1.381 mm.

Now,

$$\Delta W = \Delta L\left(1.5 - \frac{W}{2L}\right) = 1.381\,\text{mm} \tag{3.16}$$

Now,

$$G_r = \frac{W^2}{90\lambda_0^2} \tag{3.17}$$

=0.001 as W = 30 mm which is smaller than 0.35 λ_0.

$$\varepsilon_{r,n} = \frac{\varepsilon_{reff}+1}{2} = 1.165 \tag{3.18}$$

and

$$Z_r = \frac{120\pi\left[\frac{W}{h} + 1.393 + 0.667\ln\left(\frac{W}{h} + 1.444\right)\right]^{-1}}{\sqrt{\varepsilon_{r,n}}} \tag{3.19}$$

=13.41 Ω.

Q_r, Q_d, Q_c and Q_T can be calculated as

$$Q_r = \frac{\pi}{4G_r Z_r} = 53.76 \tag{3.20}$$

$$Q_d = \frac{\pi(\varepsilon_r - 1)\sqrt{\varepsilon_{r,n}}}{27.3(\varepsilon_{r,n} - 1)\sqrt{2\varepsilon_{r,n} - 1}} \frac{1}{tan\delta} = 232.19 \tag{3.21}$$

$$Q_c = h\sqrt{\pi f \mu_0 \sigma} = 1333.31 \tag{3.22}$$

$$Q_T = 42.27 \tag{3.23}$$

Therefore, the resonant resistance at edge ($x_0 = 0.5\ L$)

$$R_r = \frac{4h}{\pi\lambda_0}\mu\eta_0 Q_T\left(\frac{L + 2\Delta L}{W + 2\Delta W}\right) cos^2\left(\frac{\pi(0.5L - x_0)}{L + 2\Delta L}\right) = 333.43\Omega.$$

(b) To obtain the optimum feed point, we need to find the point where input impedance of the patch becomes 50 Ω. From part (a), we get

$L + 2\Delta L = 32.762$ mm

$W + 2\Delta W = 32.762$ mm

$Q_T = 42.27$

$R_r = 33.43\ \Omega$

$\lambda_g = 95.84$ mm

Putting these in Eq. (3.3), we may write

$$\frac{4h}{\pi\lambda_0}\mu\eta_0 Q_T\left(\frac{L + 2\Delta L}{W + 2\Delta W}\right) cos^2\left(\frac{\pi(0.5L - x_0)}{L + 2\Delta L}\right) = 50 \tag{3.24}$$

or, $(0.5L - x_0) = 12.23$

or, $x_0 = 2.77$ mm.

4. Radiation characteristics

4.1. Gain enhancement

Any RMPA has a strong influence of substrate permittivity (ε_r) on its gain. In general, an RMPA experiences a decrease in gain with the increase of ε_r. Normally, an RMPA's gain is around 6 dBi when PTFE ($\varepsilon_r = 2.33$) is used as substrate. Here, we have discussed the role of air substrate on the gain enhancement issue of a simple RMPA. It is also found in an earlier work of one of the present authors in [41] that the use of air substrate leads to a symmetrical radiation pattern of an RMPA in its two principal planes.

The gain of an RMPA loaded with air substrate is directly related to its effective radiating area A_{eff}. Therefore, we can simply compare the change of the gain of an RMPA loaded with air substrate to a conventional (reference) RMPA loaded with PTFE (**Figure 5**) as discussed in an earlier work of one of the present authors in [41]:

$$\Delta G\, in\, dB = 10\log_{10}\left[\frac{\left(A_{eff}/\lambda_0^2\right)_{air}}{\left(A_{eff}/\lambda_0^2\right)_{ref}}\right] \tag{4.1}$$

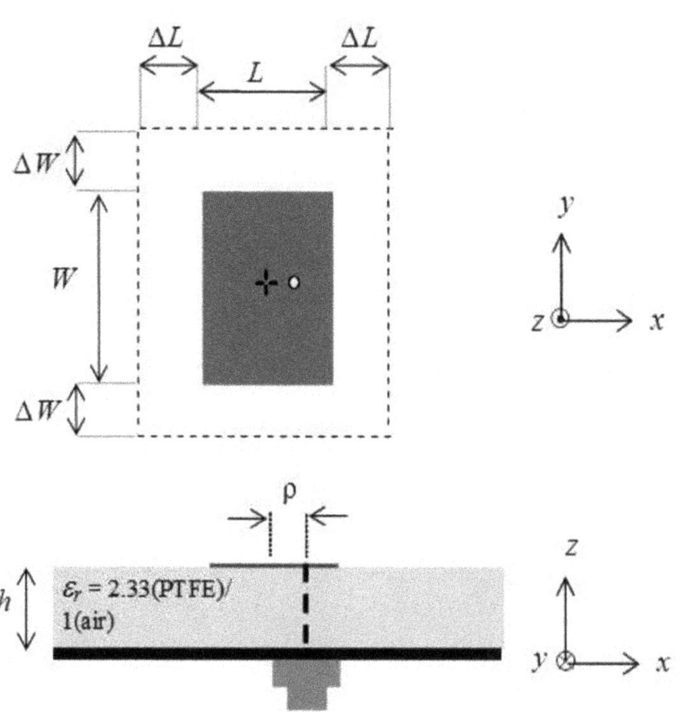

Figure 5. Schematic diagram of a rectangular patch using conventional PTFE (ε_r = 2.33) or air substrate (ε_r = 1): top and cross-sectional views.

where λ_0 is the operating wavelength. For a rectangular patch, as shown in **Figure 5**, its effective radiating area (A_{eff}) may be calculated as given in an earlier work of one of the present authors in [41]:

$$A_{eff} = (L + 2\Delta L)(W + 2\Delta W) \tag{4.2}$$

where L and W are the physical length and width of the corresponding patch, respectively. The quantities ΔL and ΔW represent the effective increments in respective dimensions caused by the fringing electric fields (discussed in Section 2.1).

Now, the standard formula of gain (G) of any rectangular aperture is (as given in one of the earlier works of the authors in [42])

$$G = \frac{4\pi A_{eff}}{\lambda^2} = \varepsilon_{ap}\frac{4\pi A_p}{\lambda^2} \tag{4.3}$$

where A_p is the physical aperture, respectively, and ε_{ap} is the aperture efficiency. Following Eq. (4.3) we can write the expression for gain of an RMPA from our earlier work in [42]

$$G = \frac{4\pi(L + 2\Delta L)(W + 2\Delta W)}{\lambda^2} \tag{4.4}$$

It is seen that when air substrate is used in lieu of PTFE substrate, the electric field lines along the patch edges becomes more relaxed or loosely bound resulting in an increase in ΔL and ΔW, and hence A_{eff} increases. Therefore, gain increases.

The formulations presented in this section are well validated against simulations and measurements [36, 41]. These formulations are found to be very much accurate for L-Ku band and for wide range of aspect ratios. **Figure 6** shows increase in gain when PTFE substrate is replaced by air substrate for *W/L* ratio 1.5.

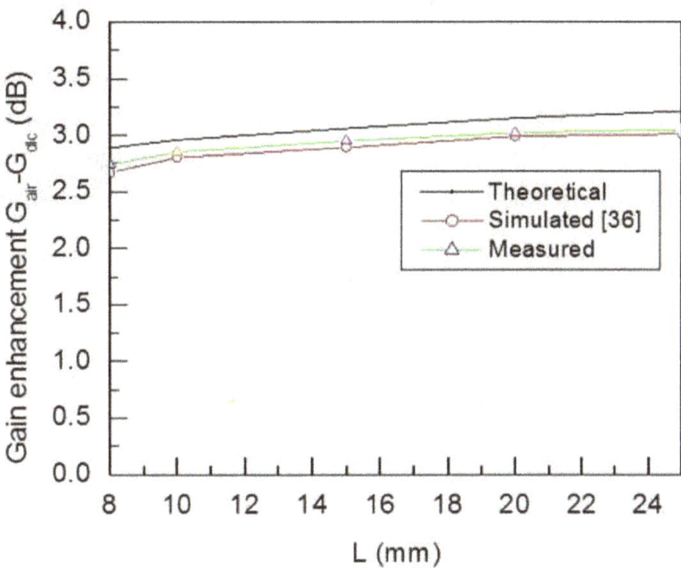

Figure 6. Variation of the gain enhancement between the microstrip patches with air and PTFE substrate for different sets of patches having the most common aspect ratio (W/L = 1.5) for ε_r = 2.33.

Theoretically computed, simulated and measurement results show very close agreement among themselves.

4.2. Polarization purity

In general, a conventional RMPA radiates in the fundamental TM_{10} mode along the broadside of the element, and the field is primarily linearly polarized, called co-polarized (CO) radiation. However, some orthogonally polarized, called cross polarized (XP), radiations take place due to weak oscillations of higher-order modes inside an RMPA. The XP radiation becomes considerably prominent for probe-fed designs particularly when the thicknesses as well as the dielectric constant of the substrate increase. Thus, the XP radiation becomes an important issue for investigation for microstrip antenna research. The (XP) fields are more significant in H plane than in E plane as obtained in our earlier work in [45]. Therefore, the polarization purity (CO-XP isolation) deteriorates in H plane (only 9 dB), and the suppression of XP radiation performance of an RMPA to improve its polarization purity is the challenging issue for antenna research community. Lower polarization purity also limits the use of RMPA in different array applications There are several techniques to improve polarization purity of an RMPA such as the use of defected ground structure (DGS) [11], grounding the non-radiating edges of a patch [43, 45] and defected patch surface [46]. A thorough discussion on DGS-integrated RMPAs can be found in [11, 42]. However, DGS-integrated RMPAs always possess high back radiation, and only 15–20 dB of CO-XP isolation in H plane can be obtained from those [11, 42]. The two later techniques can address the limitations of DGS and minimum 25 dB of CO-XP isolation from those, and these are discussed clearly in this section. The two techniques are very simple to understand and very effective to implement over a wide microwave frequency range (L-Ku band).

An RMPA with three pairs of shorting plates placed at the non-radiating edges is shown in **Figure 7**. If the non-radiating edges are grounded using pairs of thin strips, the EM boundary conditions get altered and result in a significant change in field structure with in the cavity. Hence, the restructuring of field structure within the patch inevitably modifies the radiation properties of the RMPA. Some recent work in [43, 44] show the XP radiations are typically from the non-radiating edges of the RMPA. In fact, the oscillations of electric field beneath the patch in a direction, orthogonal to E plane, produce higher-order orthogonal resonance (higher-order orthogonal modes). The XP radiations are typically due to those higher-order orthogonal modes and the fields of those modes, located near the non-radiating edges of the patch. The electric field vectors for a grounded patch in TM_{np} mode may be obtained from our earlier work in [45] as

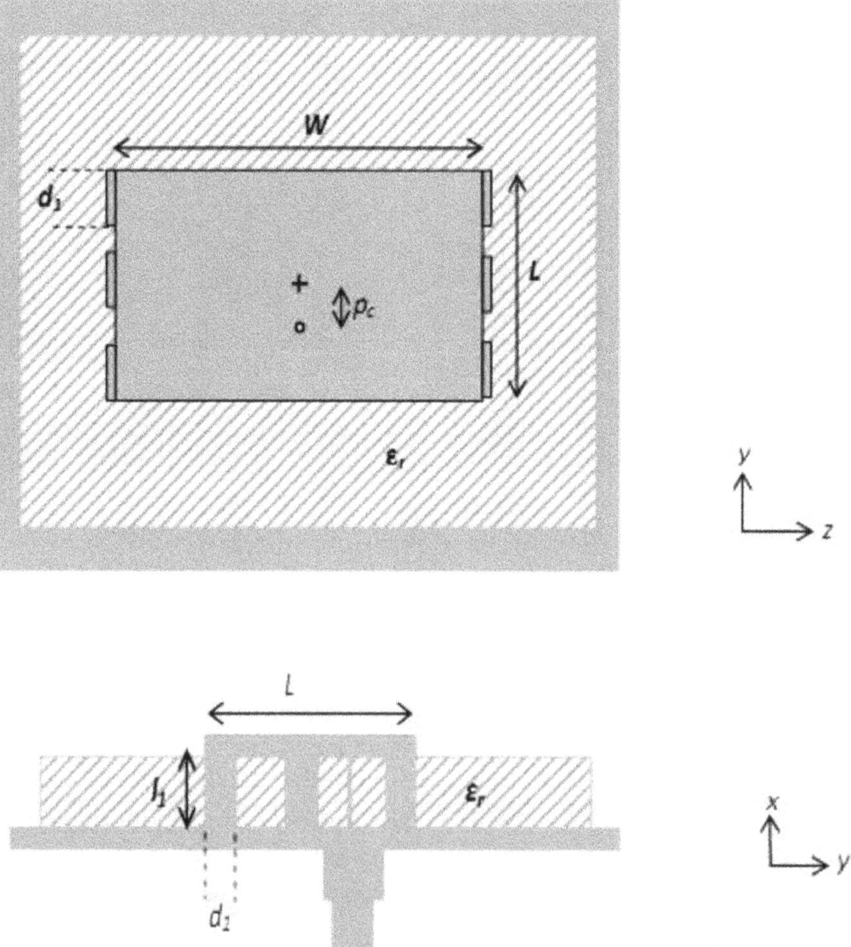

Figure 7. Schematic diagram of an RMPA with shorting strips. (a) Top view and (b) side view (L and W are the length and width of the patch, l_1 and d_1 are length and width of the shorting plates, ε_r is the dielectric constant of the substrate and p_c is the feed position).

$$E_x \propto \cos\left(\frac{n\pi}{L}y'\right)\sin\left(\frac{p\pi}{W}z'\right) \tag{4.5a}$$

$$H_y \propto \cos\left(\frac{n\pi}{L}y'\right)\cos\left(\frac{p\pi}{W}z'\right) \tag{4.5b}$$

$$H_z \propto \cos\left(\frac{n\pi}{L}y'\right)\sin\left(\frac{p\pi}{W}z'\right) \tag{4.5c}$$

where E_x, H_y and H_z are the electric and magnetic field components of the dominant mode. The number of half-wave variations along the length and width of the patch is denoted by n and p, respectively.

The electric surface current over the patch surface can be obtained from our earlier work in [45] as

$$J_s = \hat{n} \times H \tag{4.6}$$

The co-sinusoidal variation in Eq. (4.5) shows the variation of E_x along the length (L) of the patch, while the sinusoidal variation shows the variation of E_x along the orthogonal direction. Therefore, any higher-order orthogonal resonance (i.e. for any non-zero value of p) leads to minimum electric field intensity when z approaches W. In fact, the intensity of the electric fields near the non-radiating edges due to all higher-order orthogonal modes (primarily responsible for XP radiation) is forced to be minimum in order to mitigate the possibility of XP radiation from non-radiating edges due to higher-order orthogonal modes [45]. It is also seen that when the non-radiating edges are grounded with thin metallic strips, the electric fields near the non-radiating edges have least intensity when compared with the electric field intensity at the central region, but the dominant mode radiation characteristics remain unaltered. From the literature it is seen that the XP radiation from RMPA is typically from the non-radiating edges, and those are due to asymmetric field distribution along the length of the patch [43, 45]. This asymmetry in the field distribution is mainly due to the placement of feeding probe and is unavoidable for probe-fed patch. This asymmetry in the field causes asymmetry in the electric surface current (J_s) along the patch length (as discussed in our previous work in [45]). For a conventional RMPA, the y component of J_s does not become maximum at the centre [45]. This y component of J_s attributes for high XP radiation from RMPA. However, in our earlier work in [45], the use of grounded strips in case of an RMPA shows a change of the field structure beneath the patch as a result of which the electric surface current at non-radiating edge does not follow the conventional profile. Therefore, XP radiations are mitigated keeping CO radiation unaltered.

The use of grounded strip loading in a conventional RMPA not only modifies the radiation property but also regulates the input characteristics of the RMPA [45]. The length (l_1) of each thin grounding strip is essentially same as substrate thickness h, and when it is in the order of $\lambda/10$, one can write as discussed in our previous work in [45]:

$$l_1 = \lambda_{gr}/10 \tag{4.7}$$

$$\lambda_{gr} = \lambda_r/\sqrt{\varepsilon_r} \tag{4.8}$$

$$\lambda_r = c/f_r \tag{4.9}$$

where f_r is dominant mode resonant frequency, λ_{gr} is the resonant wavelength within dielectric and ε_r is the dielectric constant of substrate material.

The width of the strips (d_1) is considered to be very thin. Each grounded strip or short dipole ($l_1 \times d_1$) produces the reactive impedance [45, 47] as

$$X_s = 30\left\{2Si(kl_1) + \cos(kl_1)[2Si(kl_1) - Si(2kl_1)] - sin(kl_1)\left[2Ci(kl_1) - Ci(2kl_1) - Ci\left(\frac{2kd_1'^2}{l_1}\right)\right]\right\} \tag{4.10}$$

which comes parallel to patch input impedance (Z_p). Here, $d_1'^2$ is the equivalent circular radius of dipole of width d_1. For the dipole of noncircular cross section, $d_1'^2$ (as discussed in our previous work in [45])

$$d_1'^2 = 0.25d_1 \tag{4.11}$$

Therefore, from [45]

$$kl_1 = \frac{2\pi}{\lambda_g}\frac{\lambda_{gr}}{10} = 0.628\frac{f}{f_r} \tag{4.12}$$

and

$$\frac{2kd_1'^2}{l_1} = \frac{7.85ff_r d_1^2 \varepsilon_r}{c^2} \tag{4.13}$$

The expression for reactive impedance can be written as [45]

$$X_s = 30\left\{2Si\left(0.628\frac{f}{f_r}\right) + \cos\left(0.628\frac{f}{f_r}\right)\left[2Si\left(0.628\frac{f}{f_r}\right) - Si\left(2\times 0.628\frac{f}{f_r}\right)\right]\right.$$

$$\left. - \sin\left(0.628\frac{f}{f_r}\right)\left[2Ci\left(0.628\frac{f}{f_r}\right) - Ci\left(2\times 0.628\frac{f}{f_r}\right) - Ci\left(\frac{7.85ff_r d_1^2 \varepsilon_r}{c^2}\right)\right]\right\}. \tag{4.14}$$

The input impedance of conventional probe-fed RMPA can be written as

$$Z_p = \frac{1}{(1/R_r) + j\omega C + (1/j\omega L)} \tag{4.15}$$

where R_r is the resonant resistance of the patch at particular feed position. C and L are capacitance and inductance, respectively, and can be obtained from [9]. When this conventional probe-fed RMPA is loaded with short dipoles, the dipole reactance (X_s) will come in parallel to patch, and the resultant input impedance of dipole loaded patch can be written as [45, 47]

$$Z_{dp} = \frac{1}{\left(1/R_r\right) + j\omega C + \left(1/j\omega L\right) + jX_s} \tag{4.16}$$

Putting the values of C and L from [9]

$$Z_{dp} = \frac{1}{\left(1/R_r\right) + j\left[\frac{fQ_T}{f_r R_r} - \frac{f_r Q_T}{f R_r} + jX_s\right]} \tag{4.17}$$

where Q_T is the total quality factor.

Hence,

$$Re\left(Z_{dp}\right) = \frac{1/R_r}{\left(1/R_r\right)2 + j\left[\frac{fQ_T}{f_r R_r} - \frac{f_r Q_T}{f R_r} + jX_s\right]^2} \tag{4.18}$$

$$Im\left(Z_{dp}\right) = X_f - \frac{\left[\frac{fQ_T}{f_r R_r} - \frac{f_r Q_T}{f R_r} + jX_s\right]}{\left(1/R_r\right)^2 + j\left[\frac{fQ_T}{f_r R_r} - \frac{f_r Q_T}{f R_r} + jX_s\right]^2} \tag{4.19}$$

where X_f is the feed reactance and can be obtained from [36].

From our earlier work in [45], it is observed that when grounded strips are placed along the non-radiating edges, the structure becomes thick dipole loaded (dipole length to diameter ratio ~ 4.2 around resonant frequency), and it prevents the usual sharp variation of input reactance over the operating bandwidth. It is found from our previous work in [45] that the reactance of thick dipole slowly varies with the frequency, and as it is in parallel to patch reactance, the resultant reactance of the proposed patch varies slowly with frequency (**Figure 8**).

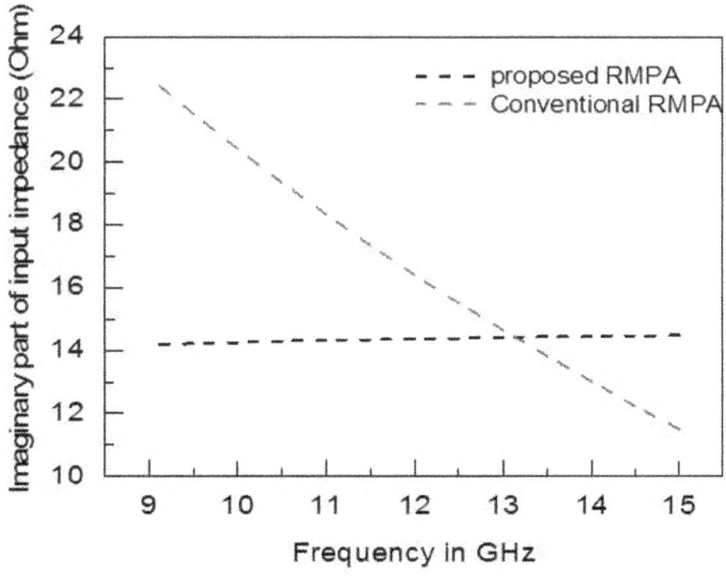

Figure 8. Variations of imaginary part of input impedance (input reactance) as a function of frequency for conventional and proposed RMPA. (Total quality factor Q_T = 0.22, resonant resistance R_r = 50 Ω and L = 8 mm, W = 12 mm, h = 1.575 mm, ε_r = 2.33, feed reactance X_f = 15 Ω). Reproduced with permission of © 2016 FREQUENZ [45].

The formulations presented in this section were validated in case of an RMPA with length L = 8 mm and W = 12 mm designed over Taconic's TLY-3-0620 PTFE material (ε_r = 2.33) with thickness h = 1.575 mm [45]. The ground plane dimensions were taken as 80 × 80 mm^2.

Three pairs of thin copper strips of thickness 0.1 mm with height of $h = l_1$ = 1.575 mm and width d_1 = 1.5 mm have been placed along the non-radiating edges (**Figure 7**). Around 25–40 dB of minimum CO-XP isolation in H plane along with input impedance of 1.32 GHz is found from the patch (see **Figure 9**) [45].

In E plane, CO-XP isolation is more than 35 dB (**Figure 10**).

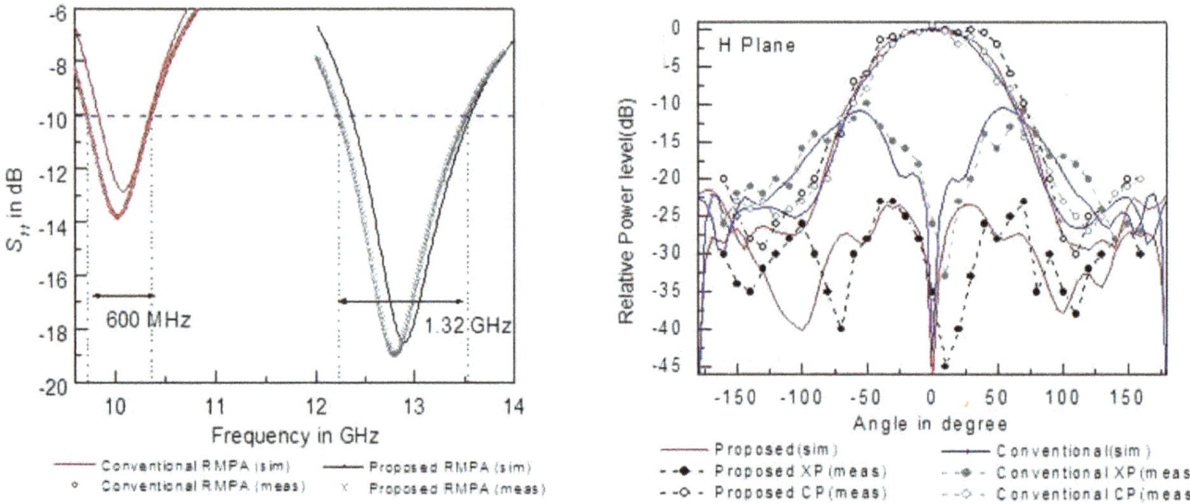

Figure 9. Simulation and measured reflection coefficient profile for conventional and proposed RMPA and simulated and measured radiation patterns for conventional (f = 10.05 GHz) and proposed RMPA (f = 12.9 GHz) at fundamental resonant mode in H plane. Reproduced with permission of © 2016 FREQUENZ [45].

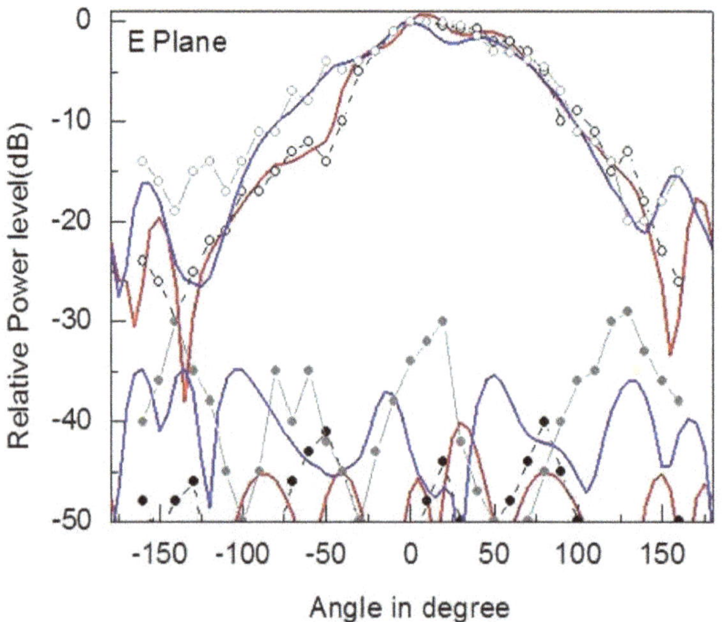

Figure 10. Comparison of simulated and measured radiation patterns for conventional (f = 10.05 GHz) and proposed RMPA (f = 12.9 GHz) at fundamental resonant mode in E plane. Reproduced with permission of © 2016 FREQUENZ [45].

One can write the field components corresponding to TM_{02} mode beneath the patch as discussed in our previous work in] (see **Figure 11** for coordinates) [46]:

$$E_x = C\cos\left(\frac{2\pi}{W}z\right) \tag{4.20}$$

$$H_y = C\sin\left(\frac{2\pi}{W}z\right) \tag{4.21}$$

Eqs. (4.20 and 4.21) show that at z = 0 and at $z = W/2$, E_x is maximum and is equal to C.

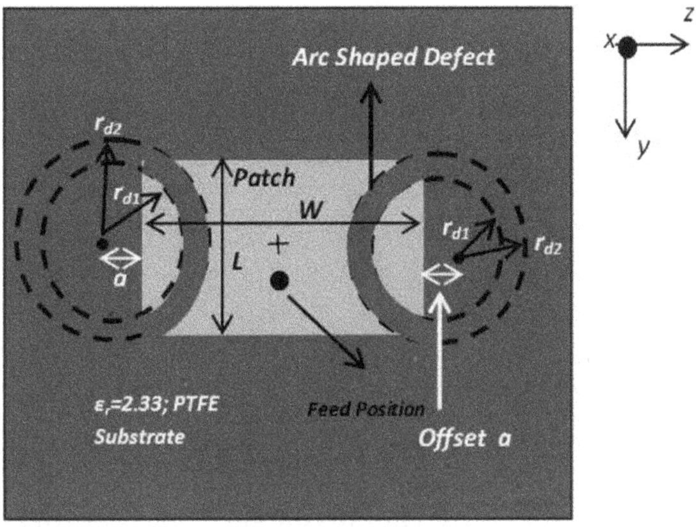

Figure 11. Schematic diagram of an RMPA with arc-defected patch surface (*top view*).

Hence, the null occurs in between these limits, i.e. [46]

$$E_x = C\cos\left(\frac{2\pi}{W}z\right) = 0 \text{ for } 0 < z < W/2 \tag{4.22}$$

From Eq. (4.22) one can write that

$$\frac{2\pi}{W}z_1 = \frac{m\pi}{2} \text{ where m} = 1, 3, 5, \ldots \tag{4.23}$$

Therefore,

$$z_1 = \frac{mW}{4} = \frac{W}{4} \tag{4.24}$$

Along the middle section of the patch, i.e. when $W \to 0$; , 97% of the maximum field exists (as discussed in our earlier work in [46]) if

$$E_x = C\cos\left(\frac{2\pi}{W}z\right) = 0.97C \tag{4.25}$$

Hence, from our previous work in [46], we can write

$$\frac{2\pi}{W}z_2 = \cos^{-1}(0.97) = \frac{2\pi}{25} \tag{4.26}$$

Therefore,

$$z_2 = \frac{W}{25} \tag{4.27}$$

A defect can be incorporated within this region, i.e. from z_1 to z_2. This in fact will perturb the fields corresponding to TM_{02} mode which is mainly responsible for XP radiation.

The electric surface current (J_s) on the patch surface corresponding to TM_{02} mode can be written as given in our earlier work in [46]:

$$J_s = \hat{n} \times H = \hat{a_x} \times |H_y| \hat{a_y} \tag{4.28}$$

Therefore,

$$J_s = \left| C sin \frac{2\pi}{W} z \right| \hat{a_z} \tag{4.29}$$

In case of an RMPA with length $L = 8$ mm and width $W = 12$ mm, the value of z_1 and z_2 is found to be 3 mm and 0.5 mm [46]. Based on the formulations described here, an arc-shaped defect has been cut on patch ($L = 8$ mm and $W = 12$ mm) symmetrically along the non-radiating edges as shown in **Figure 11** in our earlier work in [46]. The centre of the arc defect (a) is chosen in such a way that the defect can be cut through the patch corners. Around 25–35 dB, CO-XP isolation is reported from such an RMPA in H plane with such arc-shaped defect over the patch surface in [46]. However, the bandwidth is comparable with that of a conventional RMPA. Therefore, it can be observed that the effect of TM_{02} mode can be mitigated to a large extent by opting either grounding the non-radiating edges of an RMPA or judicious incorporation of defects over the patch surface. Such types of new antenna structures are surely be utmost useful for scientific and research community for designing low-profile wireless communication devices where polarization purity is required to establish over the whole operating band.

5. Conclusion

In this chapter, some recent developments in the CAD techniques have been presented lucidly but thoroughly for rectangular microstrip antennas. The presented formulations are very much accurate and are valid for wide range of aspect ratios and substrate thickness compared to other formulations. It is hoped that the work would be helpful for researchers and engineers working in the field of microstrip antennas and will help them to gain an insight into the physics of any RMPA.

Acknowledgements

Authors would like to express their deep sense of gratitude to Prof. L. Lolit Kumar Singh of Mizoram University, Mizoram; Prof. Gautam Das of Siliguri Institute of Technology, West Bengal; Prof. Debatosh Guha and Dr. Jawad Y. Siddiqui of the Institute of Radio Physics and Electronics, Calcutta University and Prof. B. N. Basu of Sir. J. C. Bose School of Engineering, Mankundu, West Bengal, India, for fruitful discussions during the preparation of the manuscript.

Subhradeep Chakraborty thanks Prof. Santanu Chaudhury, Director, CSIR-CEERI, Pilani; Dr. S. N. Joshi, Ex-Emeritus Scientist; Chirag P. Mistry, Scientist, TWT Group; Dr. Amitavo Roy Choudhury, Senior Scientist, TWT Group; Dr. Sanjay Kumar Ghosh, Principal Scientist and Head of TWT Group and Dr. R. K. Sharma, Principal Scientist and Head of MWT Division, CSIR-CEERI, Pilani, for always encouraging research endeavours and their support.

Author details

Sudipta Chattopadhyay[1*] and Subhradeep Chakraborty[2]

*Address all correspondence to: sudipta_tutun@yahoo.co.in

1 Department of Electronics and Communication Engineering, Mizoram University, Aizawl, Mizoram, India

2 TWT Group, MWT Division, CSIR-Central Electronics Engineering Research Institute, Pilani, Rajasthan, India

References

[1] Deschamp G., Sichak W., Microstrip microwave antennas. In: Third symposium on the USAF antenna research and development program; October 18, 1953; October 22, 1953. USA:1953.

[2] Howell J.Q., Microstrip antennas. IEEE Trans. Antennas Propag. 1972;**23**:90–93. DOI: 10.1109/TAP.1975.1141009.

[3] Munson R.E., Conformal microstrip antennas and microstrip phased arrays. IEEE Trans. Antennas Propag., 1974;**25**:74–78. DOI: 10.1109/TAP.1974.1140723.

[4] Bahl I.J., Bhartia P., Microstrip Antennas. USA: Artech House; 1980.

[5] Lee K.F., Luk K.M., Microstrip Patch Antennas. UK: Imperial College Press; 2011.

[6] James J.R., Hall P.S., Wood C., Microstrip Antennas: Theory and Design. London: Peter Peregrinus; 1981.

[7] James J.R., Hall P.S., editors. Handbook of Microstrip Antennas. London: Peter Peregrinus; 1989.

[8] Lee K.F., Chen W., editors. Advances in Microstrip and Printed Antennas. New York: JohnWiley& Sons Ltd; 1997.

[9] Garg R., Bhal I., Bhartia P., Ittpiboon A., Microstrip Antenna Design Handbook. Boston, USA: Artech House; 2001.

[10] Volakis J., editors. Antenna Engineering Handbook. New York: McGraw-Hill; 2007.

[11] Guha D., Antar Y.M.M., editors. Microstrip and Printed Antennas: New Trends, Techniques and Applications. UK: Wiley; 2011.

[12] Pozar D.M., Schaubert D.H., editors. Microstrip Antennas, IEEE Press, New York:1995.

[13] Gupta K.C., Benella A., editors. Microstrip Antenna Design. USA: Artech House; 1988.

[14] Chattopadhyay S., Biswas M., Siddiqui J.Y., Guha D., Rectangular microstrips with variable air gap and varying aspect ratio: improved formulations and experiments. Microw. Opt. Technol. Lett. 2009;**51**:169–173. DOI: 10.1002/mop.24025.

[15] Hammerstad E.O., Equations for Microstrip Circuit Design. In: 5th European Microwav. Conf; Hamburg:1975. pp. 268–272.

[16] Garg R., Long S.A., Resonant frequency of electrically thick rectangular microstrip antennas. Electron. Lett. 1987;**23**:1149–1151. DOI: 10.1049/el:19870801.

[17] Abboud F., Damiano J.P., Papiernik A., Simple model for the input impedance of coax-fed rectangular microstrip patch antenna for CAD. IEE Proc. Microw. Antennas Propag. 1988;**135**:323–326. DOI: 10.1049/ip-h-2.1988.0066.

[18] Thouroude D., Himdi M., Daniel J.P., CAD oriented cavity model for rectangular patches. Electron. Lett. 1990;**26**:842–844. DOI: 10.1049/el:19900552.

[19] Pues H., Vande Capelle A., Accurate transmission line model for the rectangular microstrip antenna. IEE Proc. 1984:334–340. DOI: 10.1049/ip-h-1.1984.0071.

[20] Bhattacharya A.K., Long rectangular patch antenna with a single feed. IEEE Trans. Antennas Propag. 1990;**38**:987–993. DOI: 10.1109/8.55609.

[21] Newman E.H., Tulyathan P., Analysis of microstrip antennas using moment methods. IEEE Trans. Antennas Propag. 1981;**29**:47–53. DOI: 10.1109/TAP.1981.1142532.

[22] Ridgers G.M., Odendaal J.W., Joubert J., Entire-domain versus subdomain attachment modes for the spectral-domain method of moments analysis of probe-fed microstrip patch antennas. IEEE Trans. Antennas Propag. 2004;**52**:1616–1620. DOI: 10.1109/TAP.2004.829401.

[23] Chew W.C., Liu Q., Resonance frequency of a rectangular microstrip patch. IEEE Trans. Antennas Propag. 1988;**36**:1045–1056. DOI: 10.1109/8.7216.

[24] Chang E., Long S.A., Richards W.F., Experimental investigation of Electrically thick rectangular microstrip antennas. IEEE Trans. Antennas Propag. 1986;**34**:767–772. DOI: 10.1109/TAP.1986.1143890.

[25] Kara M., Design consideration for rectangular microstrip antenna elements with various substrate thickness. Microw. Opt. Technol. Lett. 1998;**19**:111–121. DOI: 10.1002/(SICI) 1098-2760(19981005).

[26] Lee K.F., Ho K.Y., Dahele J.S., Circular disc microstrip antenna with an air-gap. IEEE Trans. on Antennas Propag. 1984;**32**:880–884. DOI: 10.1109/TAP.1984.1143428.

[27] Guha D., Resonant frequency of circular microstrip antennas with and without air-gaps. IEEE Trans. Antennas Propag. 2001;**49**:55–59. DOI: 10.1109/8.910530.

[28] Abboud F., Damino J.P., Papiernik A., Accurate Model for The Input Impedance of Coax-Fed Rctangular Microstrip Antenna With and Without Air-Gaps. In: International Conference on Antennas and Propagation; 1989. pp. 102–106.

[29] Fan Z., Lee K.F., Spectral domain analysis of rectangular microstrip antennas with an air-gap. Microw. Opt.Technol. Lett. 1992; **5**:315–318. DOI: 10.1002/mop.4650050708.

[30] Qiu J., Huang Y., Wang A., An Improved Model for the Resonant Frequency of Tunable Rectangular Microstrip Antenna. In: International Conference on Microwave and Milimeter Wave Technology; 2002. pp. 524–527.

[31] Fortaki T., Khedrouche D., Bouttout F., Benghalia A., A numerically efficient full-wave analysis of a tunable rectangular microstrip patch. Int. J. Electron. 2004;**91**:57–70. DOI: 10.1080/00207210310001656097.

[32] Zhong S.S., Liu G., Quasim G., Closed form expression for resonant frequency of rectangular patch antennas with multi dielectric layers. IEEE Trans. Antennas Propag. 1994;**42**:1360–1363. DOI: 10.1109/8.318667.

[33] Schaubert D., Pozar D., Adrian A., Effect of microstrip antenna substrate thickness and permittivity: comparison of theories and experiment. IEEE Trans. Antennas Propag. 1989;**37**:677–682. DOI: 10.1109/8.29353.

[34] Gauthier G.P., Courtay A., Rebeiz G.M., Microstrip antennas on synthesized low dielectric constant substrates. IEEE Trans. Antennas Propag. 1997;**45**:1310–1314. DOI: 10.1109/8.611252.

[35] Chattopadhyay S., Theoretical and Experimental Studies of Some Aspects of a Rectangular Microstrip Patch Antenna [Thesis]. Kolkata: University of Calcutta; 2011.

[36] HFSS: High Frequency Structure Simulator, Ansoft Corp.version 11, USA.

[37] Guha D., Siddiqui J.Y., Resonant frequency of circular microstrip antenna covered with dielectric superstrate. IEEE Trans. Antennas Propag. 2003;**51**:1649–1652. DOI: 10.1109/ TAP.2003.813620.

[38] Wolff I., Knoppik N., Rectangular and circular microstrip disk capacitors and resonators. IEEE Trans. Microw. Theory Tech. 1974;**22**:857–864. DOI: 10.1109/TMTT.1974.1128364.

[39] Guha D., Antar Y.M.M., Siddiqui J.Y., Biswas M., Resonant resistance of probe and microstrip line-fed circular microstrip patches. IEE Proc. Microw. Antennas Propag. 2005;**152**:481–484. DOI: 10.1049/ip-map:20045161.

[40] Chattopadhyay S., Biswas M., Siddiqui J.Y., Guha D. Input impedance of probe fed rectangular microstrip antenna with air gap and aspect ratio. IET Microw. Antennas Propag. 2009;**3**:1151–1156. DOI: 10.1049/iet-map.2008.0320.

[41] Guha D., Chattopadhyay S., Siddiqui J.Y., Estimation of gain enhancement replacing PTFE by air substrate in a microstrip patch antenna. IEEE Antennas. Propag. Mag. 2010;**52**:92–95. DOI: 10.1109/MAP.2010.5586581.

[42] Chakraborty S., Chattopadhyay S., Substrate fields modulation with defected ground structure: a key to realize high gain, wideband microstrip antenna with improved polarization purity in principal and diagonal planes. Int. J. RF and Microw. Computer Aided Eng. (RFMiCAE), Wiley, USA. 2016;**26**:174–181. DOI: 10.1002/mmce.

[43] Ghosh D., Ghosh S.K., Chattopadhyay S., Nandi S., Chakraborty D., Anand R., Raj R., Ghosh A., Physical and quantitative analysis of compact rectangular microstrip antenna with shorted non-radiating edges for reduced cross-polarized radiation using modified cavity model. IEEE Antennas Propag. Mag. 2014;**56**:61–72. DOI: 10.1109/LAWP.2014. 2363563.

[44] Ghosh A., Ghosh D., Chattopadhyay S., Singh L.L.K., Rectangular microstrip antenna on slot type defected ground for reduced cross polarized radiation. IEEE Antennas Wirel. Propag. Lett. 2014;**14**:321–324. DOI: 10.1109/LAWP.2014.2363563.

[45] Poddar R.P., Chakraborty S., Chattopadhyay S., Improved cross polarization and broad impedance bandwidth from simple single element shorted rectangular microstrip patch: theory and experiment. FREQUENZ. 2016;DOI: 10.1515/freq-2015-0105.

[46] Shivnarayan S.S., Vishvakarma B.R., Analysis of slot loaded rectangular microstrip patch antenna, Indian J. Radio Space Phys., 2005; **34**:424–430.

[47] Balanis, C.A., Antenna Theory: Analysis and Design. 2nd ed. Wiley; USA, 2001.

8

Metamaterial Antennas for Wireless Communications Transceivers

Mohammad Alibakhshikenari,
Mohammad Naser-Moghadasi,
Ramazan Ali Sadeghzadeh, Bal Singh Virdee and
Ernesto Limiti

Abstract

Limited space is given to antennas in modern portable wireless systems, which means that antennas need to be small in size and compact structures. However, shrinkage of conventional antennas leads to performance degradation and complex mechanical assembly. Therefore, the design of miniature antennas for application in wireless communication systems is highly challenging using traditional means. In this chapter, it is shown that metamaterial (MTM) technology offers a solution to synthesize antennas with a small footprint with the added advantage of low cost and excellent radiation characteristics.

Keywords: antennas, ultra-wideband, metamaterials, composite right-/left-hand transmission lines, microstrip

1. Introduction

In this chapter, novel and compact planar antennas are presented including ultra-wideband (UWB) and hexa-band antennas. The UWB antenna presented in part 2 is based on metamaterial (MTM) transmission lines. In reality, the MTM antenna structure is more accurately described as a composite right-/left-handed transmission line (CRLH-TL) structure due to the resulting parasitic capacitance and inductance effects. The design of the UWB antenna is achieved by embedding E-shaped dielectric slits in the radiating patches. It is shown that the dimensions of such patches have a considerable influence on the radiation characteristics of the antenna. Parametric study is undertaken to demonstrate how the MTM unit-cell's parameters affect the antennas performance in terms of gain, radiation efficiency, and radiation patterns. A novel antenna structure referred to as 'hexa-band' coplanar waveguide (CPW)-fed antenna, which is presented in part 3, consists of asymmetric fork-shaped-radiating elements incorporating U-shaped radiators with a slit. Each of the branched radiators generates triple resonant frequencies within the L-, S-, C- and X-bands.

2. Ultra-wideband antenna

Geometry of an innovative ultra-wideband (UWB) antenna in **Figure 1** shows that it is essentially a patch antenna with a rectangular slot. The antenna is terminated on the right-hand (RH) side with a grounded 50-Ω resistive load in order to reduce any impedance mismatch and thereby enhance the antenna's performance. Enclosed inside the slot are three radiating patches on which is etched an E-shaped dielectric slit. The middle patch of the antenna is excited through a coplanar waveguide (CPW) transmission line. A larger E-shaped slit is included in the antenna next to the 50-Ω load, which exhibits characteristics of the simplified composite right-/left-handed transmission line (SCRLH-TL) [1]. In the equivalent circuit of the antenna, the left-handed capacitance (C_L) results from the dielectric gap between the patches and the conductor next to the larger E-shaped slit. The impedance bandwidth of the antenna is affected by the smaller E-shaped slits and the dielectric gap represented by (C_L). In fact, the radiation properties of the antenna are affected by the dimensions of the three patches. Prototype of the antenna was fabricated on RT/duroid® RO4003 dielectric substrate, with a thickness (h) of 0.8 mm, dielectric constant (ε_r) of 3.38 and tanδ of 22×10^{-4}. The antenna design was optimized using full-wave electromagnetic (EM) simulators, that is, of Agilent Advanced Design System (ADS), High-Frequency Structure Simulator (HFSS™) and CST Microwave Studio (CST-MWS). The simulation results were validated with the actual measurements.

The configuration of the E-shaped slit was determined through simulation analysis. As it will be shown later, the size of the three patches actually has an influence of the antenna gain and efficiency properties. The equivalent circuit of the E-shaped slit with the absence of left-handed inductance (L_L) and its microstrip structure is shown in **Figure 2(a)** and **(b)**, respectively. Parameters of the simplified CRLH structure in **Figure 2** are given by

$$L(nH) = 0.2l\left[In\left(\frac{l}{W+l}\right) + 1.193 + \frac{W+l}{3l}\right] \times \left(0.57 - 0.145ln\frac{W}{h}\right) \tag{1a}$$

$$C(pF) = 10\,l\left[\frac{\sqrt{\varepsilon_e}}{Z_0} - \frac{\varepsilon_r W}{360\,\pi h}\right] \tag{1b}$$

$$\varepsilon_e = \frac{\varepsilon_r + 1}{2} + \frac{\varepsilon_r - 1}{2}\sqrt{\frac{1}{(1+10h/W)}} \tag{2}$$

where W and l represent the width and length of the microstrip line, respectively, and h and ε_r the thickness and relative permittivity of the substrate, respectively.

The E-shaped slit structure employed here is shown in **Figure 2(b)**. Based on the resonance frequencies of this structure, the initial magnitude of C_L, C_R, and L_R can be obtained from the dispersion relation of the SCRLH-TL. This is determined by applying the periodic boundary conditions related to the Bloch-Floquet theorem, given by

$$\varnothing(\omega) = \beta(\omega)p = cos^{-1}(1 + 1/ZY) \tag{3}$$

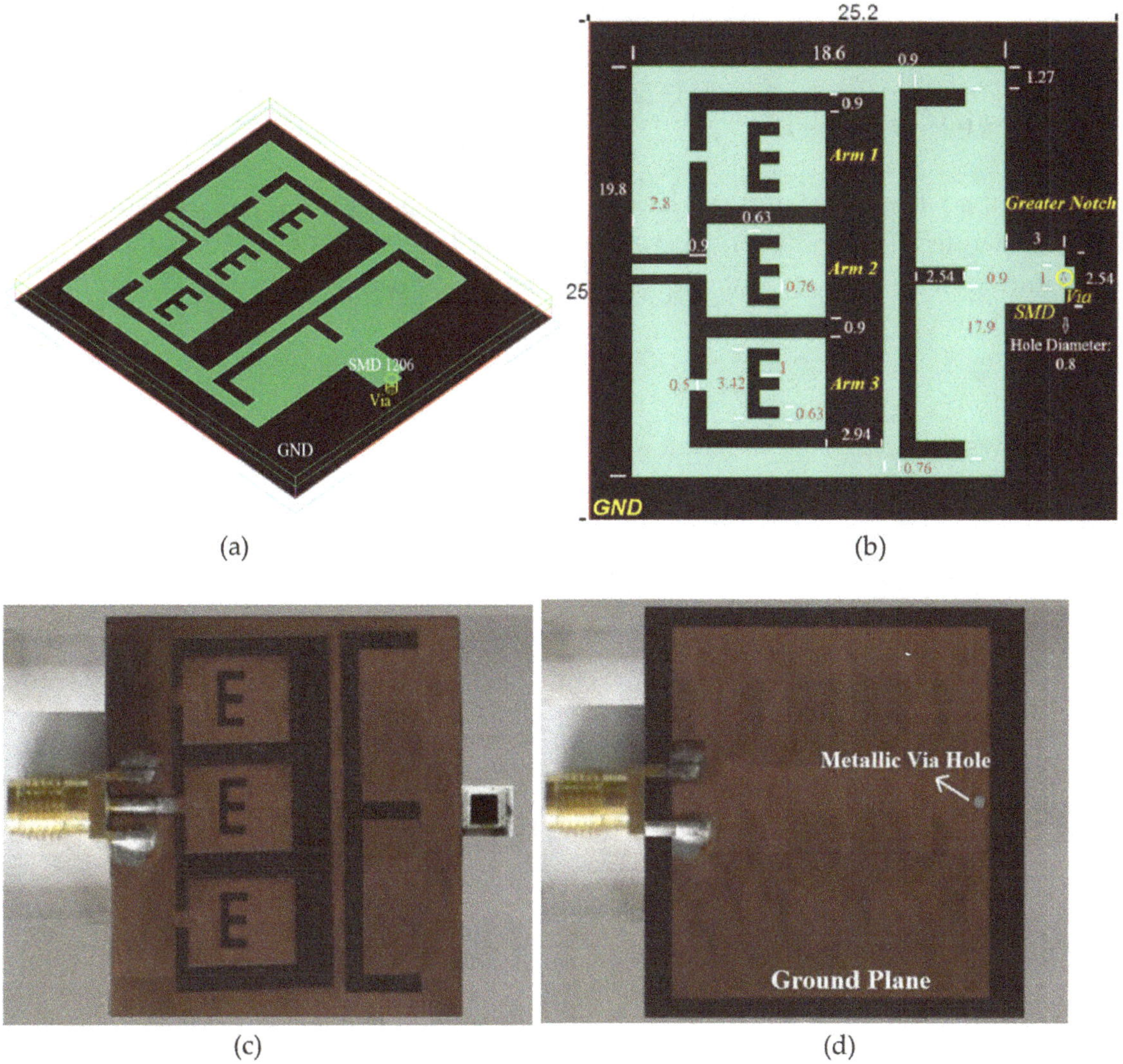

(a) (b)

(c) (d)

Figure 1. Test antenna prototype: (a) isometric view, (b) top view with dimensions annotated in millimetres, (c) top view of fabricated antenna and (d) bottom view of fabricated antenna.

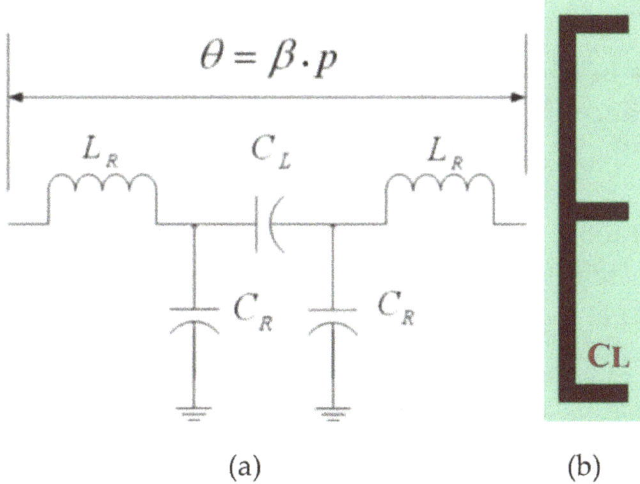

(a) (b)

Figure 2. Simplified CRLH structure without inductance L_L: (a) lumped structure and (b) distributed equivalent one.

$$Z(\omega) = j(\omega L_R - 1/\omega C_L) \tag{4}$$

$$Y(\omega) = j\left(\omega C_R - 1/\omega L_L\right) \xrightarrow{\text{SCRLH-TLwithout}C_L\text{or}L_{L,}\ \text{Here, without}L_L} Y(\omega) = j\omega C_R \tag{5}$$

The series impedance of the unit cell is represented by Z and its shunt admittance by Y. The series right-handed (RH) inductance is represented by L_R, and its shunt RH capacitance by C_R. By substituting Eqs. (4) and (5) into Eq. (3), the dispersion can be represented by

$$\varnothing(\omega) = cos^{-1}\left[1 - \left(\frac{C_R}{2C_L}\left(\omega^2 C_L L_R - 1\right)\right)\right] \tag{6}$$

Figure 3 shows the dispersion diagram of the antenna based on SCRLH-TL as determined by HFSS using Eq. (6).

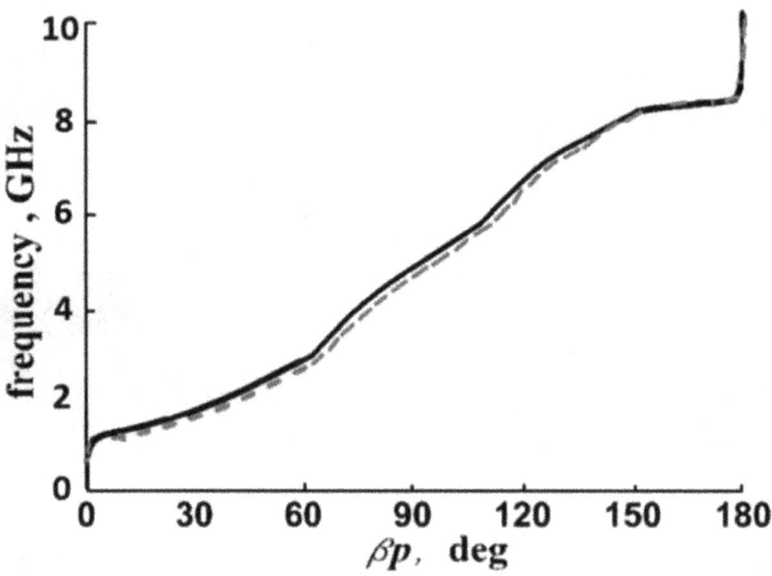

Figure 3. Dispersion diagram of the proposed antenna constructed using SCRLH-TL. HFSS™ (solid line) and Eq. (6) (dashed line).

Negative-order modes in the SCRLH-TL structure vanish with the absence of left-handed capacitance or inductance. Hence, the electrical size of the SCRLH-TL structure can be reduced by increasing the magnitude of C_L, L_R and C_R. This can be achieved by adjusting the structure dimensions. An optimization technique has been adopted within ADS, HFSS™ and CST-MWS commercial tools, leading to the final size of the structure. The resulting antenna dimensions are annotated in **Figure 1**, and the corresponding values of the lumped elements are $C_L = 8.2\ pF$, $L_R = 5.8\ nH$ and $C_R = 5.1\ pF$.

The antenna was fabricated using standard manufacturing techniques, and its performance measured. The physical dimensions of the antenna are 21.6 × 19.8 × 0.8 mm^3, and the corresponding electrical dimensions at 0.7 and 8 GHz, respectively, are $504 \times 10^{-4}\ \lambda_o \times 462 \times 10^{-4}\ \lambda_o \times 18 \times 10^{-4}\ \lambda_o$ and $576 \times 10^{-3}\ \lambda_o \times 528 \times 10^{-3}\ \lambda_o \times 21 \times 10^{-3}\ \lambda_o$, where λ_o is free-space wavelength. The ground-plane size of the antenna is 25.2 × 25 mm^2.

Anechoic chamber was used to accurately characterize the antenna's performance in terms of its gain and radiation efficiency. This involved applying RF power to the antenna and measuring the EM field radiated from it in the surrounding space. Radiation efficiency was calculated by taking the ratio of the radiated power to the input power of the antenna. Standard gain comparison technique was used to measure the antenna's gain. This involved using a pre-calibrated standard gain antenna to find the absolute gain of the antenna under test.

Figure 4 shows how the vertical dielectric gap between small patches (L_{gap}), and the horizontal gap between small patches and the metallization next to the larger E-shaped slit (W_{gap}), affects the antenna's return-loss performance. It is evident from the simulation that both these dielectric gaps can have a significant effect on the impedance bandwidth of the antenna. The simulation results show that by reducing these gaps, the antenna's impedance bandwidth can be enhanced.

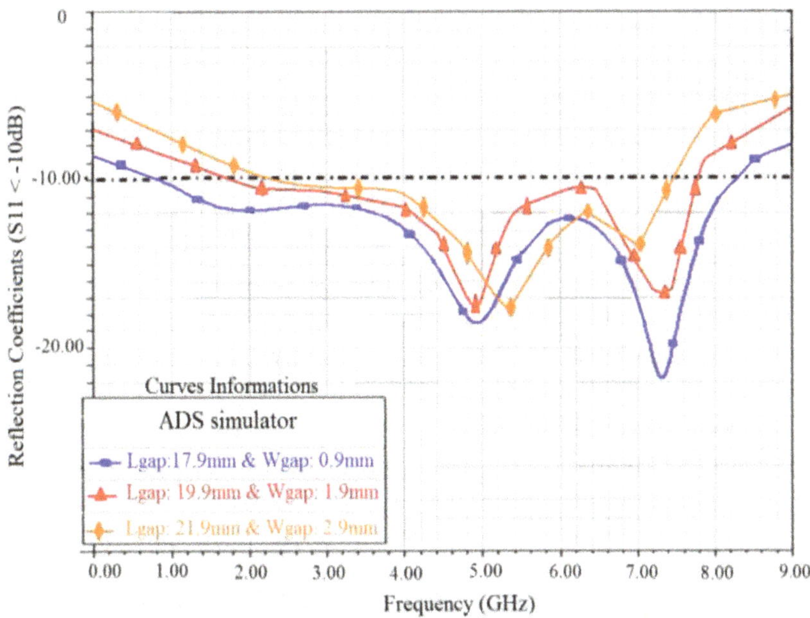

Figure 4. Effect of the gap between the three patches (L_{gap}) and the outer conductor (W_{gap}) adjacent to the larger E-shaped slit on the antenna impedance bandwidth.

Figure 5 shows the proposed antenna's measured and simulation return-loss performance, where the impedance bandwidth is defined by S_{11} ≤-10 dB. For comparison purposes, various EM simulation tools were employed to compute the impedance bandwidth of the antenna. Both HFSS™ and CST Microwave Studio (MWS) are powerful three-dimensional (3D) EM simulators, where HFSS™ is based on finite element method (FEM) technique and CST MWS is based upon finite integration technique. ADS is 2.5D EM tool based upon method of moment (MoM) technique. The impedance bandwidth predicted by the following: (1) ADS is 7.44 GHz (0.8–8.24 GHz); (2) CST MWS is 7.88 GHz (0.4–8.28 GHz) and (3) HFSS™ is 7.65 GHz (0.55– 8.2 GHz). The correlation between HFSS™ and CST-MWS is excellent. The measured fractional bandwidth is 167.8%, and the simulated fractional bandwidths are 164.6% (ADS), 181.6%, (CST-MWS) and 174.9% (HFSS™). The measured impedance bandwidth is 7.3 GHz (0.7–8 GHz). This confirms that the antenna can operate over an ultra-wideband.

The antenna resonates at two distinct frequencies, as shown in **Figure 5**, which are measured at 4.75 and 7 GHz. Simulation results predict 4.15 and 6.8 GHz (ADS), 4.03 and 6.68 GHz (CST-MWS), and 4 and 6.6 GHz (HFSS™). The measured results agree well with the ADS prediction. The divergence in the results is attributed to manufacturing tolerances.

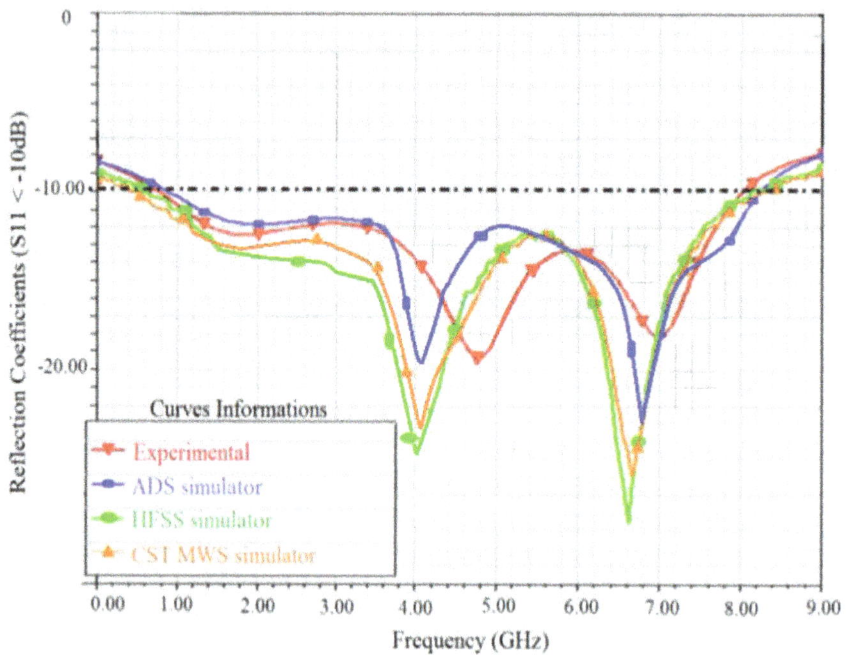

Figure 5. Antenna measured and simulated reflection coefficient response.

The distribution of the current density over the surface of the antenna at the two resonance frequencies of 4.75 and 7 GHz is shown in **Figure 6**. The current density distribution is symmetrical with reference to the feed line.

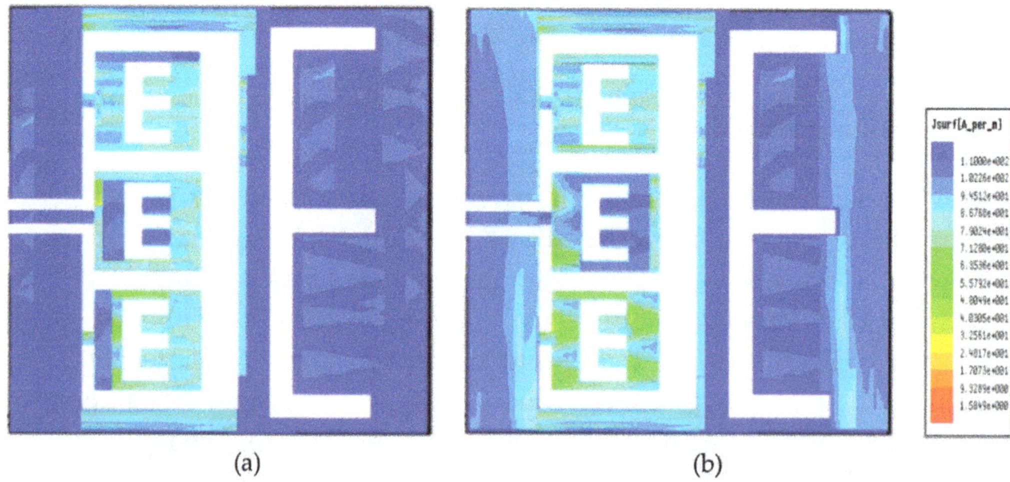

Figure 6. Current density distribution over the antenna at (a) 4.75 GHz and (b) 7 GHz.

Excitation of the proposed antenna structure though the SubMiniature version A (SMA) connector's feed line can result in an imbalance in the current flow over the outer metallization of the antenna, which can severely undermine the radiation characteristics of the antenna. To prevent this from happening, it was necessary to incorporate three smaller patches in the antenna slot. As is evident in **Figure 1**, only the middle patch is excited through the feed line, whereas the other two patches are used as parasitic radiators. This configuration facilitates the concentration of the EM fields in the proximity of the antenna structure instead of dispersing the field over its ground plane, which would otherwise contribute in unwanted coupling. **Figure 7** shows that a larger number of patches can significantly improve the antenna's bandwidth and gain properties. The antenna was implemented on a RT/duroid® RO4003 substrate with total dimensions of about 20 × 20 mm^2. Within this size, the maximum number of inner patch arms was restricted to three. The resulting antenna gain and radiation efficiency at its operating frequency range are given in **Table 1**.

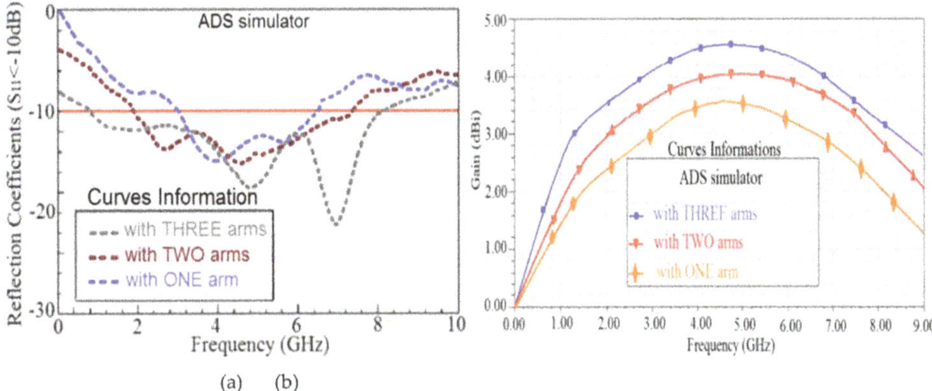

Figure 7. Simulated antenna reflection coefficient and gain response as a function of the number of inner patch arms.

Freq. (GHz)	Simulation (ADS)		Measured	
	Gain (dBi)	Efficiency (%)	Gain (dBi)	Efficiency (%)
0.70	1.8	53	1.2	50
4.75	4.6	82	4.0	80
7.00	3.7	75	3.6	73
8.00	3.1	70	3.1	68

Table 1. Antenna gain and radiation efficiency.

The radiation patterns of the antenna were simulated using ADS, CST-MWS and HFSS™. The simulated and measured radiation patterns are shown in **Figure 8** at the two resonance frequencies.

The antenna radiates similarly to a dipole antenna with a large coverage angle at the two resonance frequencies. **Figure 9** shows the simulated and measured radiation gains of the antenna. The measured radiation gain is greater than 1.2 dBi between 0.7 and 9 GHz, with a peak of 4.2 dBi at 4.8 GHz.

Table 2 shows comparison of the proposed antenna with similar antenna structures published in literature to date. It exhibits the largest fractional bandwidth and highest efficiency.

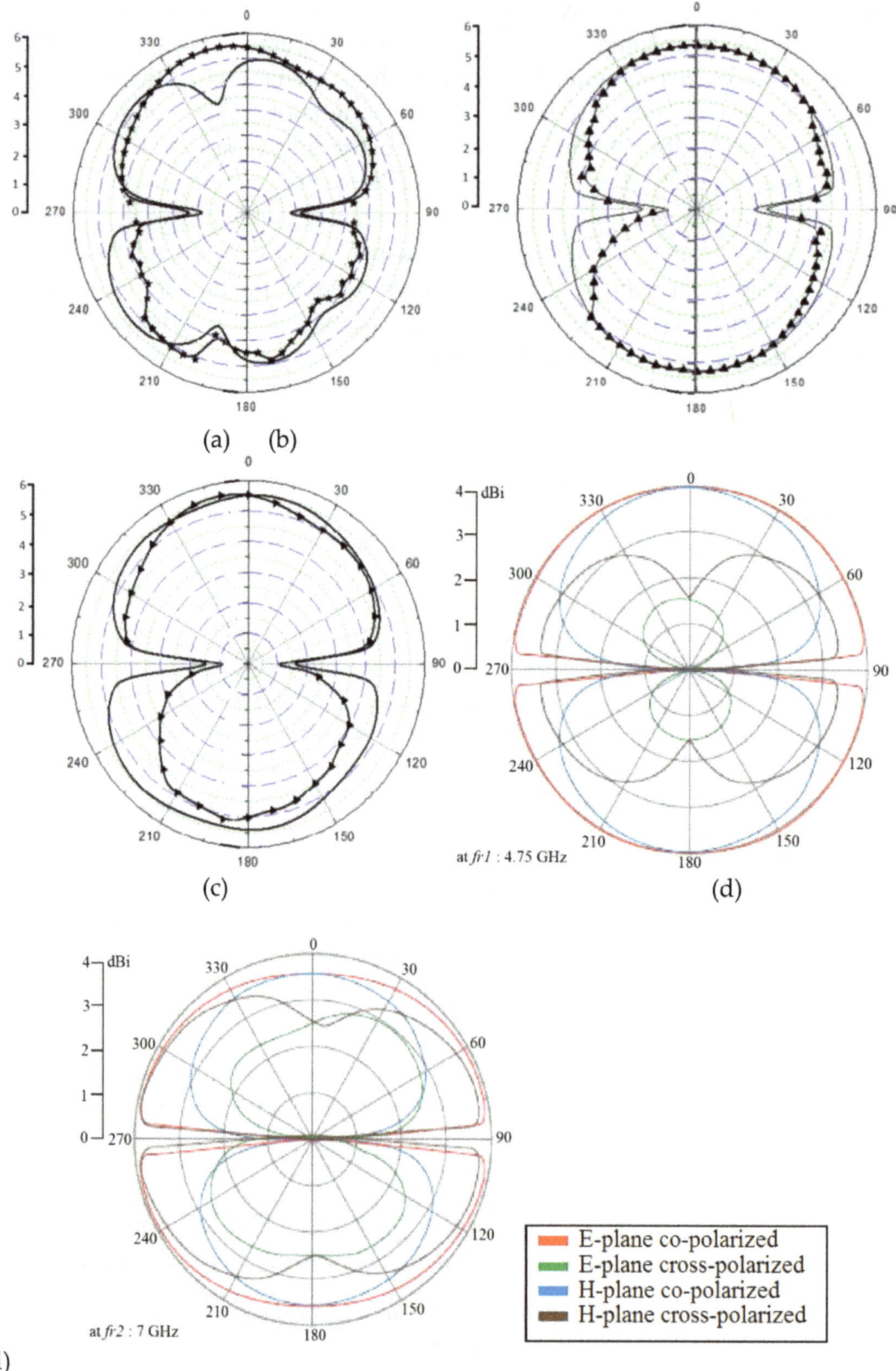

(d)

Figure 8. Simulated (E-plane) and measured (E-plane and H-plane) radiation patterns of the proposed antenna at the two resonance frequencies. (a) ADS at f_{r1} = 4.95 GHz (* line) and f_{r2} = 7.33 GHz (— line), (b) CST-MWS at f_{r1} = 4.03 GHz (▲ line) and f_{r2} = 6.68 GHz (— line), (c) HFSS at f_{r1} = 4 GHz (▲ line) and f_{r2} = 6.6 GHz (— line) and (d) measured at f_{r1} = 4.75 GHz and f_{r2} = 7 GHz.

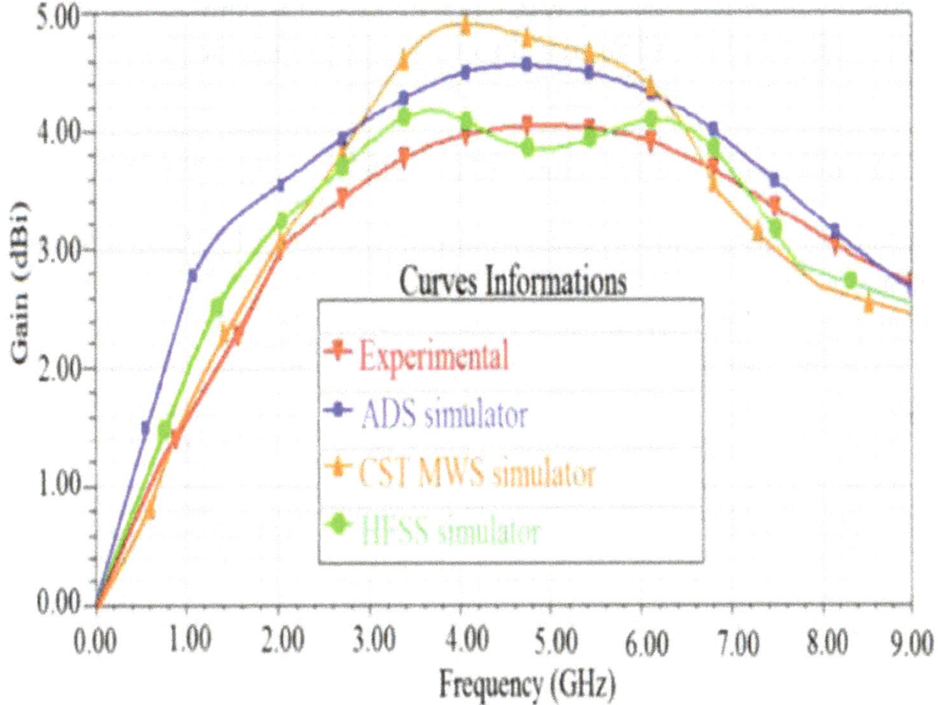

Figure 9. Antenna simulated and measured gain as a function of frequency.

Papers	Dimensions at 1 GHz	Bandwidth	Gain (max)	Eff. (max)
[2] 4 × UC antenna	$0.047\ \lambda_o \times 0.021\ \lambda_o \times 0.002\ \lambda_o$ 14.2 × 6.32 × 0.8 mm^3	104.8% (1–3.2 GHz)	2.3 dBi	62%
[2] 6 × UC antenna	$0.064\ \lambda_o \times 0.021\ \lambda_o \times 0.0027\ \lambda_o$ 19.2 × 6.32 × 0.8 mm^3	123.8% (0.8–3.4 GHz)	2.8 dBi	70%
[3] 5 × UC antenna	$0.056\ \lambda_o \times 0.02\ \lambda_o \times 0.005\lambda_o$ 16.7 × 7 × 1.6 mm^3	82.9% (7.7–18.6 GHz)	3.1 dBi	58%
[3] 6 × UC antenna	$0.06\ \lambda_o \times 0.02\ \lambda_o \times 0.027\ \lambda_o$ 18 × 7 × 0.8 mm^3	74.4% (7.5–16.8 GHz)	2.1 dBi	44%
[3] 7 × UC antenna	$0.072\ \lambda_o \times 0.02\ \lambda_o \times 0.005\lambda_o$ 21.7 × 7 × 1.6 mm^3	87.2% (7.8–19.85 GHz)	3.4 dBi	68%
[3] 8 × UC antenna	$0.075\ \lambda_o \times 0.02\ \lambda_o \times 0.027\ \lambda_o$ 22.6 × 7 × 0.8 mm^3	84.2% (7.25–17.8 GHz)	2.3 dBi	48%
[4]	$0.2\ \lambda_o \times 0.05\ \lambda_o \times 0.003\ \lambda_o$ 60 × 16 × 1 mm^3	116.7% (0.67–2.55 GHz)	4.74 dBi	62.9%
[5]	$0.06\ \lambda_o \times 0.06\ \lambda_o \times 0.005\ \lambda_o$ 18 × 18 × 1.6 mm^3	26.5% (1.8–2.35 GHz)	3.69 dBi	20%
[6]	$0.2\ \lambda_o \times 0.017\lambda_o \times 0.017\lambda_o$ 60 × 5 × 5 mm^3	103% (0.8–2.5 GHz)	0.45 dBi	53.6%
[7]	$0.06\lambda_o \times 0.06\lambda_o \times 0.021\lambda_o$ 18.2×18.2×6.5 mm^3	66.7% (1–2 GHz)	0.6 dBi	26%
[8]	$0.04\lambda_o \times 0.04\lambda_o \times 0.011\lambda_o$ 12×12×3.33 mm^3	8.2% (2.34–2.54 GHz)	1 dBi	22%
[9]	$0.07\lambda_o \times 0.08\lambda_o \times 0.003\lambda_o$ 20×25×0.8 mm^3	8.3% (3.45–3.75 GHz)	2 dBi	27%
Proposed antenna	$0.072\lambda_o \times 0.066\lambda_o \times 0.002\lambda_o$ 21.6×19.8×0.8 mm^3	167.8% (0.7–8 GHz)	4 dBi	80%

Table 2. Antenna performance compared to other miniature UWB antennas (UC: Unit Cell).

It has the second highest gain and its relatively small size is comparable to most previous works. These results clearly confirm that the antenna is great candidate for ultra-wideband communication systems.

2.1. Effect of the 50-Ω load on the antenna bandwidth

The return loss of the antenna with and without the 50-Ω load is shown in **Figure 10**. It is evident that the 50-Ω load enhances the impedance match so that the impedance bandwidth improves from 131.8% (without 50 Ω) up to 164.6% (with 50 Ω). Excellent match is discerned near the two resonance frequencies of the antenna, at 4.8 and 7 GHz, thus enabling effective radiation at the resonance frequencies. These results are summarized in **Table 3**.

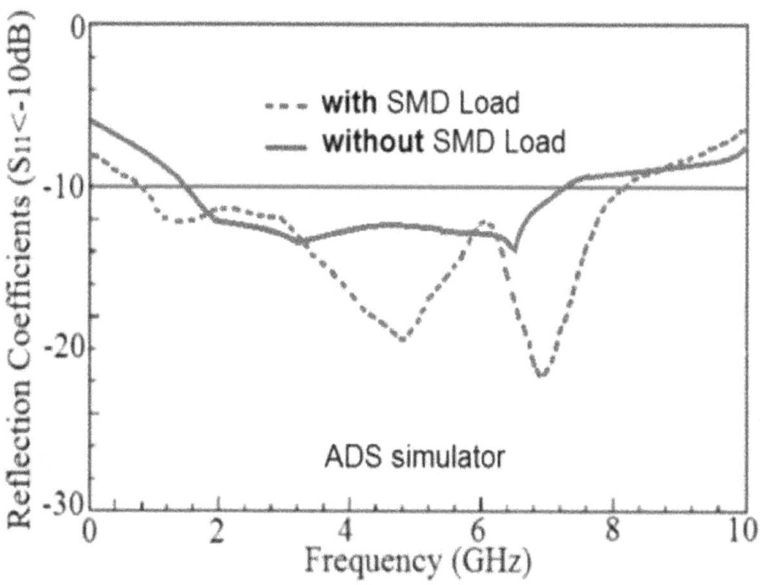

Figure 10. Antenna reflection coefficient (S_{11}) as a function of frequency with and without a 50-Ω load.

Unloaded	1.5–7.3 GHz, Δf = 5.8 GHz ≈ 131.8%
Loaded	0.8–8.24 GHz, Δf = 7.44 GHz ≈ 164.6%

Table 3. Antenna bandwidth with and without 50-Ω load.

2.2. Effect of the SMD load on the radiation performance

Antenna gain and efficiency performance, with and without the 50-Ω impedance load, are shown in **Figure 11**, and the results are tabulated in **Table 4**. These results show that a small improvement is achieved with loading the antenna.

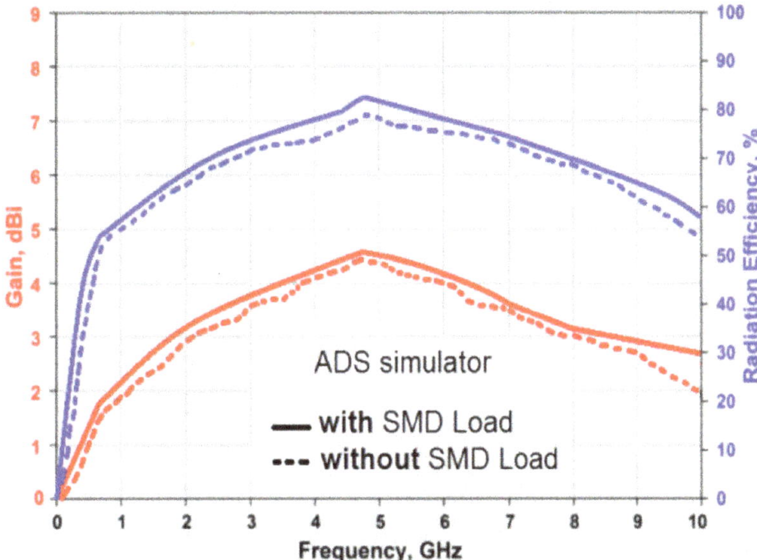

Figure 11. Antenna gain and efficiency with and without 50-Ω impedance loading simulated as a function of frequency.

Unloaded				
Frequency (GHz)	0.7	4.75	7	8
Gain (dBi)	1.6	4.5	3.6	3
Efficiency (%)	51	79	73	69
Loaded				
Frequency (GHz)	0.7	4.75	7	8
Gain (dBi)	1.8	4.6	3.7	3.1
Efficiency (%)	53	82	75	70

Table 4. Radiation properties unloaded/loaded by 50 Ω.

3. Hexa-band antenna

Configuration of a novel hexa-band CPW-fed antenna, shown in **Figure 12**, comprises three asymmetric fork-shaped-radiating stubs with U-shaped-radiating elements. Embedded within the U-shaped elements is a dielectric slit. The tri-branched radiator generates three distinct resonant frequencies within the L-, S-, C- and X-bands. The U-shaped element also generates resonant frequencies at the lower band of the antenna. The asymmetrical fork-shaped stubs improve the impedance-matching characteristics of the antenna and reduce its stopband. The antenna is fabricated on an FR4 substrate with relative permittivity $\varepsilon_r = 4.4$ and thickness $h = 1.6$ mm and is fed through a coplanar waveguide transmission line with 50-Ω impedance. The substrate size is 35 × 26 mm^2.

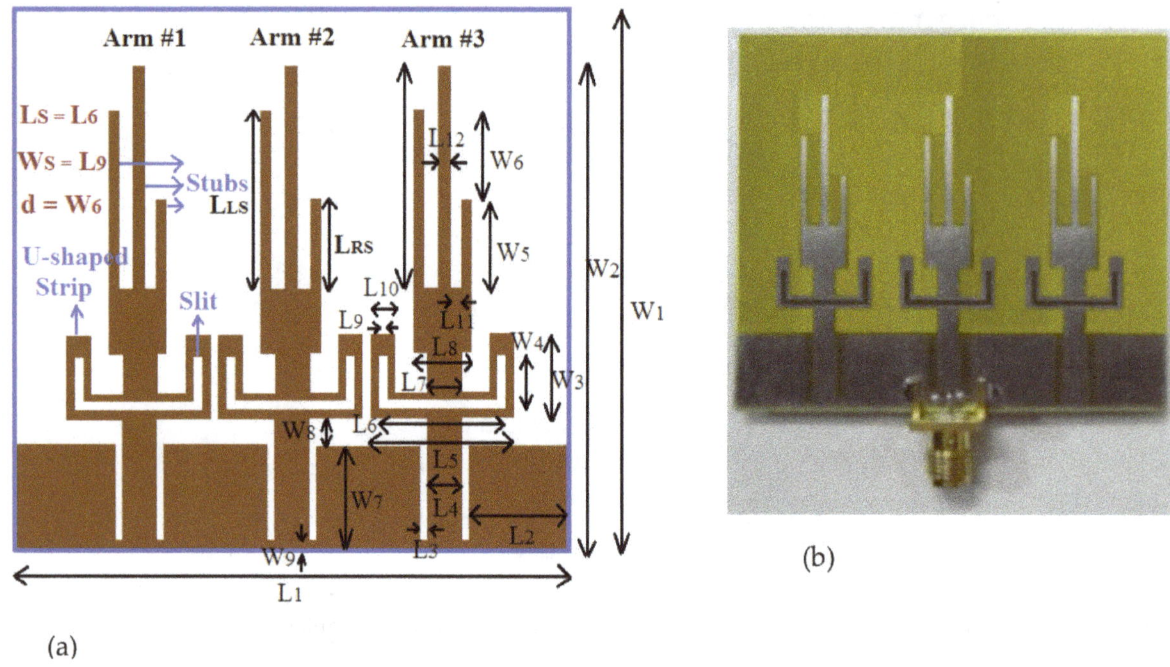

Figure 12. CPW-fed monopole hexa-band antenna: (a) top view of the proposed antenna and (b) the fabricated antenna.

The antenna was excited through a 50-Ω coplanar waveguide transmission line. **Figure 13** shows that it resonates at 1.3, 1.75, 3.35, 4.85, 6.5 and 7.6 GHz. There is good agreement between simulated and measured results. Any divergence between the simulated and measured results is ascribed to a number of factors, that is (1) the non-uniform current density distribution over the antenna structure, (ii) imperfect equivalent circuit models, (iii) undesirable EM coupling and (iv) manufacturing errors. This antenna is shown to function between 700 MHz and 11.35 GHz, and has a fractional bandwidth of 176.76% that includes several communication standards, in particular, GSM, DCS, PCS, Bluetooth, WLAN, WiMAX and WiFi along with the major parts of the C- and X-bands. The measured E-plane and H-plane radiation characteristics at 1.3, 1.75, 3.35, 4.85, 6.5 and 7.6 GHz are shown in **Figure 14**. The measured results show that the antenna radiates omnidirectionally at these frequencies.

In **Figure 13**, the impedance bandwidth of the lower passband at 1.3 GHz is 800 MHz (0.7–1.5 GHz) with a corresponding fractional bandwidth of 72.72%. The impedance bandwidth of the next passband at 1.75 GHz is 1100 MHz (1.6–2.7 GHz) with S_{11} better than 15 dB and functions at GSM (upper), DCS, PCS, WiFi (lower), Bluetooth, WiMAX (lower) and WLAN (2.4–2.484 GHz). The impedance bandwidth of the passband at 3.35 GHz is 1.75 GHz (2.85–4.6 GHz) covering the WiMAX (upper) band. The subsequent passbands at 4.85 GHz, at 6.5 GHz and at 7.6 GHz have impedance bandwidths of 1.35 GHz (4.65–6 GHz), 0.95 GHz (6.25–7.2 GHz) and 4 GHz (7.35–11.35 GHz), respectively, covering WiFi (upper) band and significant parts of C- and X-bands. The impedance bandwidth, passband and stopband of antenna-I are given in **Table 5**.

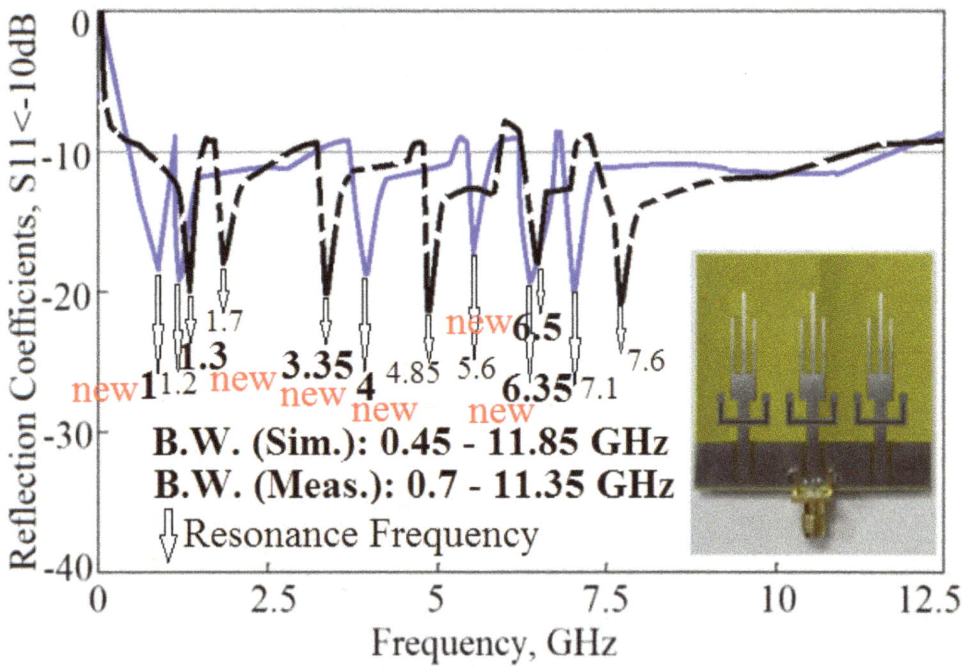

Figure 13. Simulated (solid-line) and measured (dashed-line) S_{11} response of the antenna.

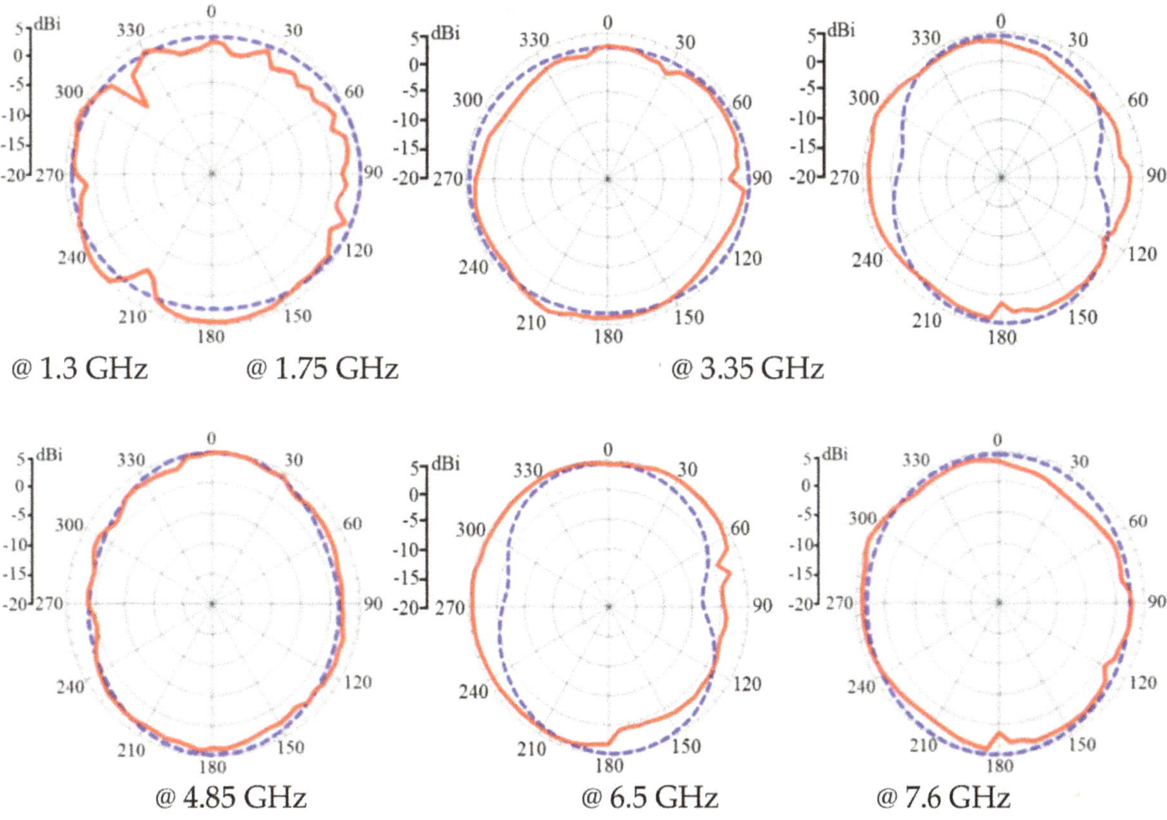

Figure 14. Measured E-plane (dashed-line) and H-plane (solid-line) radiation patterns of the antenna at the resonance frequencies.

Total Imp. BW	176.76% (0.7–11.35 GHz)
Passband 1	72.72% (0.7–1.5 GHz) → f_{r1} =1.3 GHz
Stopband 1	1.51–1.59 GHz
Passband 2	51.16% (1.6–2.7 GHz) → f_{r2} =1.75 GHz
Stopband 2	2.71–2.84 GHz
Passband 3	46.97% (2.85–4.6 GHz) → f_{r3} =3.35 GHz
Stopband 3	4.61–4.64 GHz
Passband 4	25.35% (4.65–6 GHz) → f_{r4} =4.85 GHz
Stopband 4	6.01–6.24 GHz
Passband 5	14.12% (6.25–7.2 GHz) → f_{r5} =6.5 GHz
Stopband 5	7.21–7.34 GHz
Passband 6	42.78% (7.35–11.35GHz) → f_{r6} =7.6GHz

Table 5. Impedance bandwidth of the antenna (f_{rN}: resonance frequency of band N).

Table 6 shows a comparison of the proposed antenna with similar antennas published in literature. Parameters compared are (i) the number of distinct bands, (ii) the communication standards being covered and (iii) the sizes of the antennas. It is evident that the proposed antenna operates in the following communication bands: GSM (880–960 MHz/1.85–1.99 GHz), DCS (1.71–1.88 GHz), PCS (1.71–1.99 GHz), Bluetooth (2.402–2.480 GHz), WLAN (2.4/5.2/5.8 GHz), WiMAX (2.3–2.4/2.496–2.690/3.3–3.8 GHz), WiFi (2.412–2.4835/4.9–5.9 GHz) and parts of the C- and X-bands. The antenna provides substantially greater coverage than other antennas reported in Refs. [9, 10, 14, 15]. Dimensions of the proposed antenna are comparable to other antennas.

Multiband antennas	Proposed antenna	[9]	[10]	[11]	[12]	[13]	[14]	[15]
Number of separated bands	6	3	3	3	4	3	2	2
Coverage bands								
C-band	√	–	–	–	–	–	–	√
X-band	√	–	–	–	–	–	–	–
WiMAX	√	√	–	–	√	√	–	–
Bluetooth	√	–	–	–	–	√	–	–
WLAN	√	√	–	–	√	–	√	√
WiFi	√	–	√	√	√	√	–	–
PCS	√	–	–	–	–	–	√	–
DCS	√	–	√	√	–	–	–	–
GSM	√	–	–	–	–	–	–	–
Max. gain (dBi)	5.3	3.06	2.57	3.6	5.0	6.7	5.5	5.2
Dimensions (mm^2)	35 × 26	32 × 28	25 × 38	20 × 20	44 × 56	50 × 50	54 × 52	32.5 × 25

Table 6. Characteristics of the antenna in comparison with recent work.

4. Conclusion

To summarize, a compact antenna comprising three asymmetrical branched fork with U-shaped strips is shown to exhibit hexa-band characteristics. The antenna meets multi-communication standards including L-, S-, C- and X-bands for GSM, DCS, PCS, Bluetooth, WLAN, WiMAX and WiFi applications. Slits in the U-shaped strips are shown to excite additional resonant bands. The proposed structure reduces stopbands and improves the antenna's impedance-matching performance. The antenna exhibits good return loss, gain and radiation patterns, which makes it an excellent candidate for multiband and broadband communication applications.

Acknowledgements

The Authors would like to give their special thanks to faculty of microelectronics for the financial support.

Author details

Mohammad Alibakhshikenari[1*], Mohammad Naser-Moghadasi[2], Ramazan Ali Sadeghzadeh[3], Bal Singh Virdee[4] and Ernesto Limiti[1]

*Address all correspondence to: Alibakhshikenari@ing.uniroma2.it

1 Department of Electronic Engineering, University of Rome Tor Vergata, Rome, Italy

2 Faculty of Engineering, Science and Research Branch, Islamic Azad University, Tehran, Iran

3 Faculty of Electrical Engineering, K. N. Toosi University of Technology, Tehran, Iran

4 London Metropolitan University, Center for Communications Technology, London, UK

References

[1] X.Q. Lin, R.P. Liu, X.M. Yang, J.X. Chen, X.X. Ying, Q. Cheng, "Arbitrarily dual-band components using simplified structures of conventional CRLH TLs," *IEEE Transactions on Microwave Theory and Techniques*, 2006; 54(7): 2902–2909. DOI: 10.1109/TMTT.2006.877434

[2] M. Alibakhshi-Kenari, "Printed planar patch antennas based on metamaterial," *International Journal of Electronics Letters* 2014;2(1):37–42. DOI: 10.1080/21681724.2013.874042

[3] M. Alibakhshi-Kenari, "Introducing the new wideband small plate antennas with engraved voids to form new geometries based on CRLH MTM-TLs for wireless applications", *International Journal of Microwave and Wireless Technologies*, 2014;6(06):629-637.

[4] J. Luo, S. Gong, P. Duan, C. Mou, M. Long, "Small-size wideband monopole antenna with CRLH-TL for LTE mobile phone," *Progress in Electromagnetics Research C*, 2014;50:171–179.

[5] M.A. Abdalla, A.A. Awad, K.M. Hassan, "Wide band high selective compact metamaterial antenna for 2 GHz Wireless applications," *Antennas and Propagation Conf. (LAPC)*, 2014:350–354.

[6] Y. Li, Z. Zhang, J. Zheng, Z. Feng, "Compact heptaband reconfigurable loop antenna for mobile handset," *IEEE Antennas and Wireless Propagation Letters*, 2011;10:1162–1165.DOI: 10.1109/LAWP.2011.2171311

[7] C.J. Lee, K.M.K.H. Leong, T. Itoh, "Composite right/left-handed transmission line based compact resonant antennas for RF module integration," *IEEE Transactions on Antennas Propagation*, 2006; 54(8): 2283–2291. DOI: 10.1109/TAP.2006.879199

[8] C.-C. Yu, M.-H. Huang, L.-K. Lin, Y.-T. Chang, "A compact antenna based on MTM for WiMAX," *Asia-Pacific Microwave Conference (APMC)*, 2008; 2: 1127–1130.

[9] W. Hu, Y.-Z. Yin, P. Fei, X. Yang, "Compact triband square-slot antenna with symmetrical L-strips for WLAN/WiMAX applications," *IEEE Antennas Wireless Propagation Letter*, 2011; 10: 462–465. DOI: 10.1109/LAWP.2011.2154372

[10] J. Pei, A.-G. Wang, S. Gao, W. Leng, "Miniaturized triple-band antenna with a defected ground plane for WLAN/WiMAX applications," *IEEE Antennas Wireless Propagation Letter*, 2011; 10: 298–301. DOI: 10.1109/LAWP.2011.2140090

[11] N. Amani, M. Kamyab, A. Jafargholi, A. Hosseinbeig, J.S. Meiguni, "Compact tri-band metamaterial-inspired antenna based on CRLH resonant structures," *Electronics Letters*, 2014; 50(12): 847–848. DOI: 10.1049/el.2014.0875

[12] Y.F. Cao, S.W. Cheung, T.I. Yuk, "A multiband slot antenna for GPS/WiMAX/WLAN systems," *IEEE Transactions on Antennas and Propagation*, 2015; 63(3): 959–958. DOI: 10.1109/TAP.2015.2389219

[13] V.V. Reddy, N.V.S.N. Sarma, "Triband circularly polarized Koch fractal boundary microstrip antenna," *IEEE Antennas and Wireless Propagation Letters*, 2014; 13: 1057–1060. DOI: 10.1109/LAWP.2014.2327566

[14] C. Wang, Z.-H. Yan, P. Xu, J.-B. Jiang, B. Li, "Trident-shaped dual-band CPW-fed monopole antenna for PCS/WLAN applications," *Electronics Letters*, 2011; 47(4): 231–232. DOI: 10.1049/el.2010.3250

[15] X. Yang, Y.Z. Yin, W. Hu, K. Song, "Dual-band planar monopole antenna loaded with pair of edge resonators," *Electronics Letters*, 2010; 46(21): 1419–1421. DOI: 10.1049/el.2010.8349

9

Compact Antenna with Enhanced Performances Using Artificial Meta-Surfaces

Tong Cai, He-Xiu Xu, Guang-Ming Wang and Jian-Gang Liang

Abstract

In recent years, artificial meta-surfaces, with the advantages of smaller physical space and less losses compared with three-dimensional (3D) metamaterials (MTM), have intrigued a great impetus and been applied widely to cloaks, subwavelength planar lenses, holograms, etc. Typically, one most important part for meta-surfaces' applications is to improve the performance of antennas. In this chapter, we discuss our effort in exploring novel mechanisms of enhancing the antenna bandwidth using the magneto-electro-dielectric waveguided meta-surface (MED-WG-MS), achieving circular polarization radiation through fractal meta-surface, and also realizing beam manipulation using cascaded resonator layers, which is demonstrated from aspects of theoretical analysis, numerical calculation, and experimental measurement. The numerical and measured results coincide well with each other. Note that all designed antenna and microwave devices based on compact meta-surfaces show advantages compared with the conventional cases.

Keywords: magneto-electro-dielectric waveguided meta-surface, fractal meta-surface, gradient phase meta-surface, miniaturization, beam manipulation, bandwidth enhancement

1. Introduction

Microstrip antennas have been used widely in recent wireless communication systems due to their coplanar structures, easy fabrication, and stable performances. However, conventional microstrip antennas face huge challenges, such as large size, narrow bandwidth, and inconvenient tuning of the working frequency.

Recently, artificial meta-surfaces, planar inhomogeneous metamaterials composed of carefully selected elements with specific electromagnetic (EM) responses, have attracted much attention due to their strong abilities to control the wavefront of transmitted and reflected EM waves. With the unique EM properties, meta-surfaces have found a lot of applications in focusing lens, cloaks, absorbers, antennas, and other microwave devices. One of the most important applications is to improve the performances of antenna, such as extending the working bandwidth and realizing circularly polarization radiation.

In Section 1, a new concept of planar magneto-electro-dielectric waveguided meta-surface (MED-WG-MS) is introduced for the first time, which is capable of manipulating the effective permeability μ_{eff} and the effective permittivity ε_{eff} independently. As a result, the MED-WG-MS can be tuned with a larger refractive index and larger wave impedance, which is essential to realize antenna miniaturization and bandwidth enhancement, respectively. Based on the derived principle of the MED-WG-MS, a meta-surface antenna working at 3.5 GHz is proposed by properly using the MED-WG-MS. The designed antenna, occupying an area of only 0.20 λ_0 × 0.20 λ_0, realizes a series of advantages such as a 42.53% miniaturization compared with the conventional patch antenna, a 207% impedance BW enhancement, and also comparable far-field performances. In Section 2, a novel method to design the circularly polarized antenna is proposed using a combination of fractal meta-surface and fractal resonator. We find that fractal reactive impedance surface (RIS), i.e., Hilbert-shaped RIS, can reduce the substrate thickness and improve the front-to-back ratio of the antenna. More importantly, fractal resonator, i.e., Wunderlich-shaped fractal complementary split ring resonator (WCSRR), is able to achieve the CP property and further size reduction. A CP antenna working at the Wimax band is experimentally engineered with a compact layout, a relative wide axial ratio (AR) bandwidth, and a comparable radiation gain. In Section3, an ultra-thin transmissive gradient phase meta-surface (TGMS) is proposed based on the generalized Snell's law of refraction. The polarization-controlled property of the local element is demonstrated. An ultra-thin polarization beam splitter (PBS) working at the X-band is implemented by a specially designed 2D TPGM and is launched by a wideband horn antenna from the perspective of high integration, simple structure, and low cost. The polarized splitting ratio, indicating the separation level of reflected beams, is proposed and measured to systematically evaluate the performances of PBS. Finally, we summarize the chapter simply in the last section.

2. Compact microstrip antenna with enhanced bandwidth using planar MED-WG-MS

Microstrip patch antennas are experimentally demonstrated with a lot of advantages, such as coplanar configuration, simple design, and low cost, which have widespread applications in a wireless communication system recently. However, the conventional microstrip antenna is electrically large, resulting from that the working frequency is exclusively dependent on the antenna size, which limited a further application of the patch antenna. Moreover, the antenna suffers from an intrinsically narrow bandwidth (BW). To address these issues, electromagnetic (EM) meta-surfaces (MSs) have been proposed in recent years. Due to their strong EM abilities in improving the performances of conventional devices, MS antenna has become a research hotspot with remarkable achievements. However, using the MSs to increase the impedance BW of an electrically smaller microstrip patch antenna is still rarely reported.

Artificial magneto-dielectric substrate has been verified to be a promising avenue to reduce the antenna size and extend the working band width [1–4]. A reported patch antenna, using a magneto-dielectric substrate, achieves a relative 3.2% BW ordered by 6 dB return loss. However, the antenna suffers from a high profile, which is difficult for integration in compact devices. A magneto-dielectric substrate using an embedded meander line (EML) is proposed in reference [5]. A patch antenna based on this substrate realizes an antenna miniaturization and a BW improvement. Unfortunately, it needs an additional shield metal plate that complexes the fabrication. Therefore, it is an essential issue to design an MS antenna realizing simultaneously size reduction, BW extension, and also flexible frequency turning. In this section, we explored an improved strategy to simultaneously address aforementioned issues. An electrically smaller MS element is proposed by combining the electro-dielectric and magneto-dielectric waveguided substrates, defined as the magneto-electro-dielectric waveguided MS (MED-WG-MS) [6]. The MED-WG-MS provides a freedom to control the wave impedance and refractive index simultaneously, achieving antenna miniaturization, bandwidth enhancement, and also flexible frequency modulation.

2.1. The concept and working mechanism of MED-WG-MS

As discussed in reference [5], planar WG-MS is a special kind of artificial material residing in the planar waveguide environment, which is made up of an upper metallic layer and a lower ground plane. By introducing electric and magnetic resonators in the upper and lower metallic layers, MED-WG-MS is composed, which would bring about a series of interesting characteristics, including EM manipulation, further miniaturization, and also bandwidth enhancement. In addition, the MED-WG-MS is able to tune the working frequency easily, since more freedom is provided by the element. In other words, the antenna can work at any frequency just by manipulating the electric or magnetic resonators.

Then, we will discuss about the working mechanisms of the MED-WG-MS. There is a link between the patch size and the refractive index of the used substrate for a conventional patch antenna, which can be calculated as Eq. (1a) [7], where $\varepsilon_{\mathrm{ref}}$ is the effective permittivity of the substrate. The quality factor Q is inversely proportion to the working bandwidth. The Q of a microstrip antenna is written as Eq. (1c), here G_{r} denotes the radiation conductance. For an MS antenna, with a tunable effective permeability μ_{eff} and the effective permittivity $\varepsilon_{\mathrm{eff}}$, the basic principle is almost the same as the conventional counterpart. The length and quality factor of this MS antenna can be calculated in Eqs. (1b) and (2b) [6].

$$L_{ref} \approx \frac{c}{2f_0\sqrt{\varepsilon_{ref}}} = \frac{c}{2f_0 n_{ref}} \tag{1a}$$

$$L_{MS} \approx \frac{c}{2f_0\sqrt{\varepsilon_{eff}\mu_{eff}}} = \frac{c}{2f_0 n_{eff}} \tag{1b}$$

$$Q_{ref} = \frac{\pi w_p}{4 G_r h \eta_{ref}} = \frac{\pi w_p \sqrt{\varepsilon_{ref}}}{4 G_r h \eta_0} \tag{2a}$$

$$Q_{MS} = \frac{\pi w_p}{4 G_r h \eta_{eff}} = \frac{\pi w_p \sqrt{\varepsilon_{eff}}}{4 G_r h \eta_0 \sqrt{\mu_{eff}}} \tag{2b}$$

Therefore, normalized with the conventional patch antenna, the compact factor (CF), defined as the ratio between the patch size for the conventional antenna and the MS antenna, can be derived from Eq. (1a) and (1b) as

$$CF = \frac{n^2{}_{ref}}{n^2{}_{eff}} = \frac{\varepsilon_{ref}}{\varepsilon_{eff}\mu_{eff}} \tag{3}$$

The BW improving factor (BIF), evaluated by the BW ratio between the conventional antenna and the MS antenna, can be calculated as

$$BIF = \frac{Q_{ref}}{Q_{MS}} = \frac{\eta_{eff}}{\eta_{ref}} = \frac{\sqrt{\varepsilon_{ref}\mu_{eff}}}{\sqrt{\varepsilon_{eff}}} \tag{4}$$

Based on Eqs. (1)–(4) [6], CF and BIF can be controlled by manipulating the material parameters ε_{eff} or μ_{eff}. Our goal in this section is to achieve a smaller CF and a larger BIF to improve the patch antenna performance. The reported magneto-dielectric substrate [1, 4] is just a special case for manipulating the parameter μ_{eff}, however, the proposed MED-WG-MS in this chapter can control both ε_{eff} and μ_{eff}, simultaneously.

2.2. The circuit model and EM property of MED-WG-MS

Based on the analysis in Section 1.1, we proposed a basic element as shown in **Figure 1(a)** [6]. The element is a well-known sandwich structure, consisting of an upper metallic two-turn

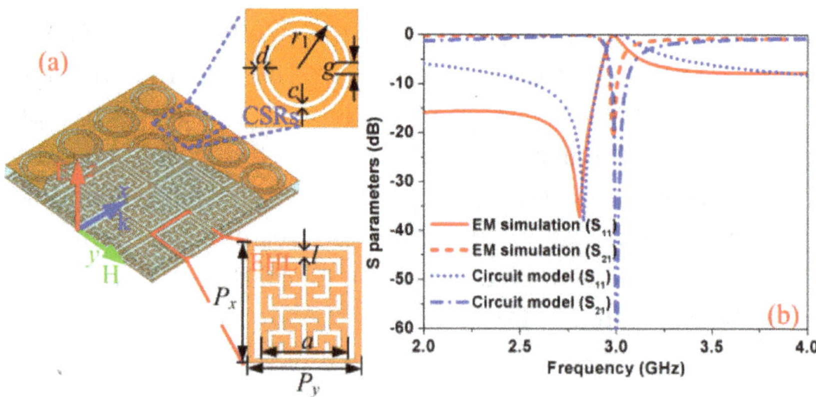

Figure 1. Schematic of the basic MED-WG-MS cell as well as the calculated *S*-parameters [6]. (a) Perspective view of the proposed element and the geometrical parameters (with the unit in mm); (b) *S*-parameters of the basic element based on the EM simulation and circuit model calculation. The detailed parameters of the CSR and EHL are $p_x = p_y = 5$ mm, $a = 3.6$ mm, $l = 0.3$ mm, $r_1 = 2.1$ mm, $d = 0.3$ mm, $c = 0.2$ mm, $g = 0.4$ mm. The circuit parameters are extracted as $C_0 = 0.11$ pF, $L_{\text{EHL}} = 17.13$ nH, $C_{\text{CSR}} = 1.21$ pF, $L = 0.34$ nH and $C = 8.09$ pF.

complementary spiral ring resonator (CSR) and a lower embedded Hilbert-line (EHL) resonator, separated by a 1.5-mm-thick F4B spacer (dielectric constant $\varepsilon_r = 2.65$ and the loss tangent $\tan\delta = 0.001$).

The meta-surface, composing of a periodic element, is shined with a plane wave as the E-field polarized along the z-direction and propagation along the x-direction, as the simulation set-up shown in **Figure 1(a)**. For a systematical analysis of the basic element, we proposed the effective circuit model shown in **Figure 2**. Under a waveguide circumstance, the conventional sandwich structure (composing of metallic layers and a dielectric spacer) can be represented by a parallel resonant tank (consisting of an inductor L and a capacitor C) and a parallel shorted capacitor C_0. The MD-WG-MS is designed by introducing the magnetic EHL in the ground plane of the basic element. A fractal EHL can be used effectively to control the ε_{eff} of the material. As a result, an additional inductor L_{EHL} is added in the parallel branch to represent the effect of the EHL as shown in **Figure 2(a)** [6]. An ED-WG-MS is composed of etching a two-turn complementary spiral ring resonator (CSR) in the upper metallic layer, as shown in **Figure 2(b)**. CSR has been demonstrated with compact size and electric resonant property in references [8, 9], which is suitable to tune the permeability μ_{eff}. And an additional capacitor C_{CSRs} is added in the series branch of the circuit model. By integrating the effect of the CSR and EHL, a new MED-WG-MS is designed with the structure shown in **Figure 2(c)**. Consequently, the circuit model can be obviously obtained as shown in **Figure 2(c)**. Further analysis shows that the permeability μ_{eff} is decided by the series branch, whereas the permittivity ε_{eff} is decided by the shunt branch. In other words, both material parameters μ_{eff} and ε_{eff} can be individually manipulated by adjusting the geometrical parameters of the EHL and CSR. For characterization, we obtain the EM performance of the element by using the finite element method (FEM)-based commercial software Ansoft HFSS, and the circuit parameter by using the Ansoft Serenade. **Figure 1(b)** illustrates the calculated S-parameters by EM simulation and circuit simulation as a function of frequency. There is a reasonable agreement between them, indicating a rationality of the circuit model. As can be seen, band gaps at about 3 GHz are clearly observed.

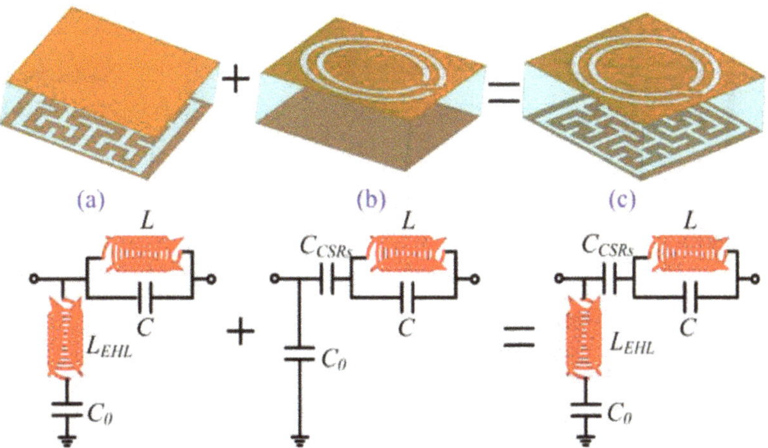

Figure 2. The evolution of WG-MS and the relative equivalent circuit model [6]. (a) MD-WG-MS, (b) ED-WG-MS, and (c) MED-WG-MS.

Next, we extracted the effective constitutive material parameters of four types of WG-MSs using the standard retrieval process [10, 11] based on the calculated *S*-parameters; the results are shown in **Figure 3**. With introduced CSR structures, electric resonances (about 2.86 GHz) are observed clearly for the MED-WG-MS, 90°-rotated MED-WG-MS and ED-WG-MS cases. However, by inserting EHL, magnetic resonances (about 2.95 GHz) can be seen for the MED-WG-MS, 90°-rotated MED-WG-MS, and MD-WG-MS structures. Routinely, the corresponding antiresonances in vicinity of these resonant frequencies are also appeared in the curves of material parameters, resulting from the magneto-electric coupling among cells and the finite size of the periodic structure [12]. The losses of these elements, represented by the imaginary part of the material parameters, can be ignored in all cases. Due to the anisotropic property of the proposed MED-WG-MS element, there is a slight resonant frequency shift and a small difference between the material parameters for the element and its 90°-rotated counterpart. A comparison of the material parameters indicates that the CSR can realize a manipulation of $\varepsilon_{\mathrm{eff}}$ and EHL accounts for the control of μ_{eff}. As is expected, according to Eqs. (3) and (4), a desirable combination of a smaller $\varepsilon_{\mathrm{eff}}$ with flat slope and a larger μ_{eff} is

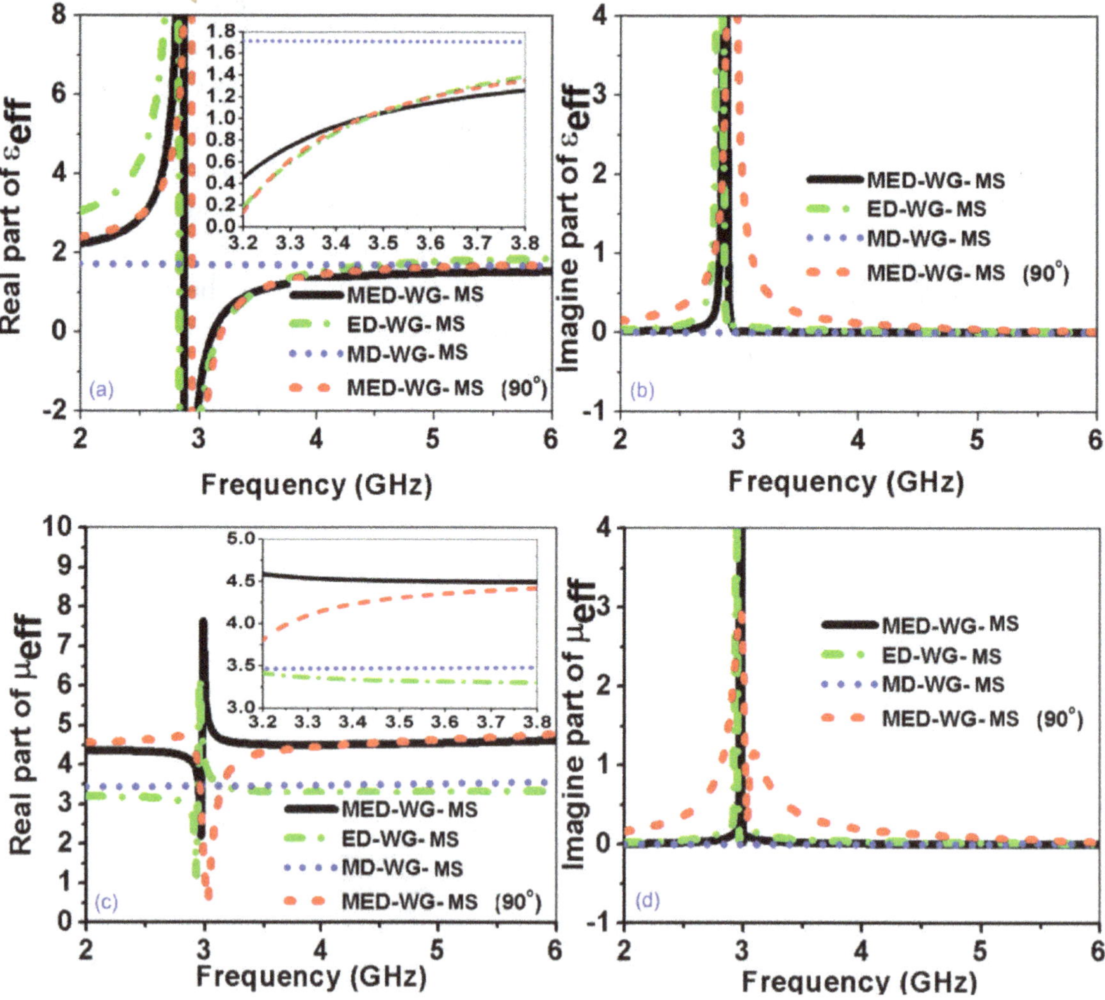

Figure 3. The effective material parameters for four different MS cells. (a) The real part and (b) imaginary part of the permittivity $\varepsilon_{\mathrm{eff}}$; (c) the real part and (d) imaginary part of the permeability μ_{eff}.

obtained for the proposed MED-WG-MS element. The good performances of the designed element indicate that it is capable of improving the antenna bandwidth and reducing the antenna size. However, undesirable small μ_{eff} and large ε_{eff} are obtained for the MD-WG-MS. While small μ_{eff} and ε_{eff} with a sharp slope are obtained for the ED-WG-MS structure. Both of the ED-WG-MS and MD-WG-MS elements are not able to improve the performance of the bandwidth and reduce the structure dimension.

2.3. Design of compact microstrip antenna with enhanced bandwidth

With a desirable MED-WG-MS element in hand, we can obtain effective parameters $\mu_{eff} = 4.5 + j0.025$ and $\varepsilon_{eff} = 1.05 + j0.033$ at 3.5 GHz, according to the aforementioned analysis. Upon introducing these effective parameters into Eqs. (3) and (4), CF = 0.41 and BIF = 2.87 can be calculated, which are fascinating for a patch antenna realizing size miniaturization and BW enhancement simultaneously. For proof of the analysis, we design and optimize a microstrip antenna working at 3.5 GHz by loading the MED-WG-MS elements. At frequency of 3.5 GHz, the radiator length of a patch antenna can be estimated as $L_{MS} \approx 20$ mm based on Eq. (1). The MED-WG-MS element occupies an area of $p_x \times p_y = 5$ mm × 5 mm, therefore, a total of 4 × 4 elements are chosen in the final antenna design, with the schematic shown in **Figure 4** [6]. In the ground plane, 8 × 6 EHL cells are stacked to maintain a constant current. Moreover, the ground just under the feed-line remains unetched to guarantee the continuity of the input energy.

Frequency tuning is essential in determining the antenna performances. In the following part, we will derive a three-step frequency tuning method to achieve a flexible frequency tuning property. First, a coarse control over the operating frequency is considered by changing the outer radius r_1 of the CSR structure. As r_1 varies from 1.8 to 2.2 mm in steps of 0.1 mm, **Figure 5(a)** depicts the frequency-dependent reflection coefficients [6]. To ensure a fair comparison, the residual geometrical parameters of the CSR are kept as $p_x = p_y = 5$mm, $d = 0.3$ mm,

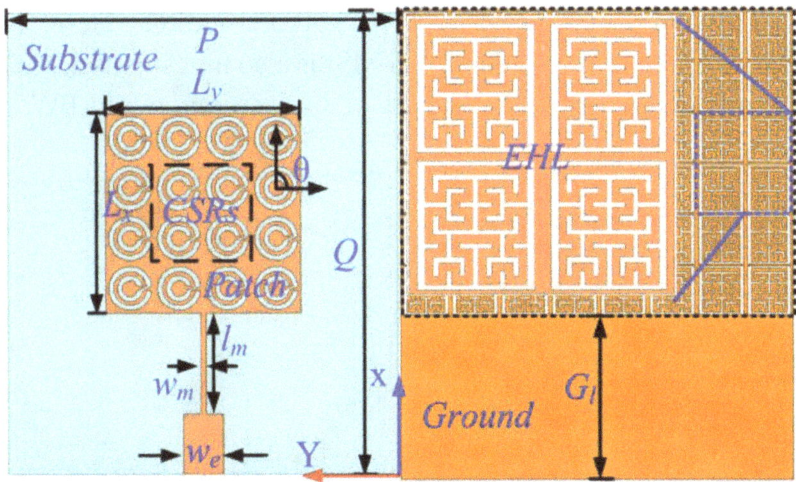

Figure 4. Schematics of the proposed antenna with MED-WG-MS loading [6]. (a) Top view; (b) bottom view. The geometrical parameters (in mm) of the proposed patch antenna are as: $P = 40$ mm, $Q = 45$ mm, $L_x = L_y = 20$ mm, $G_l = 15$ mm, $l_m = 10$ mm, $w_m = 0.8$ mm, $w_e = 4.1$ mm; the final geometrical parameters of CSRs are $r_1 = 2.1$ mm, $d = 0.5$ mm, $c = 0.3$ mm, and $g = 0.5$ mm.

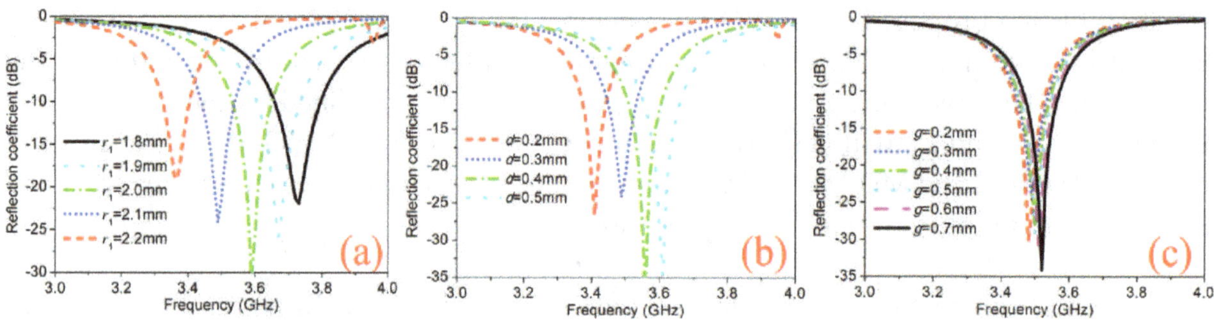

Figure 5. Simulated reflection coefficients against frequency [6]. The dependence of reflection coefficients on parameters (a) r_1, (b) d, and (c) g.

c = 0.2 mm, g = 0.4 mm. The valleys of reflection coefficients shift from 3.35 to 3.72 GHz as r_1 changes. In all cases, good impedance-matching performances are obtained for all cases with the reflection better than 19 dB. Second, a fine control over the operating frequency is studied as the parameter d increases from 0.2 to 0.5 mm. Here, the value d determines the inner circle size of CSR. Referring to **Figure 5(b)**, the reflection valleys undergo a continuous blue shift, changing from 3.4 to 3.6 GHz. Third, **Figure 5(c)** plots the reflection coefficients with a tunable parameter g. It can be seen clearly that only a small resonant frequency shift (3.47 to 3.53 GHz) appear as g varies from 0.2 to 0.7 mm. Therefore, the parameter g affords an exact control over the operating frequency. It is worth noting that the proposed three-step frequency tuning method provides a comprehensive guideline in designing antenna with an exact working frequency. Compared with the conventional patch antenna, the working frequency can be controlled easier by using the derived method.

The bianisotropic response of the CSR has a large effect on the antenna behavior. By rotating the gap orientation θ from 0 to 180°, **Figure 6(a)** plots the reflection coefficients against frequency. As θ is 90°, best performances of the antenna with wider bandwidth and lower return loss are obtained. A slightly changing resonant frequency results from the different interactions between the EHL and the CSR with different gap orientations. Here, the dependence of antenna performances on different kinds of WG-MSs is also investigated, as the results shown in **Figure 6(b)**. Desirable antenna performances of size reduction and BW enhancement are

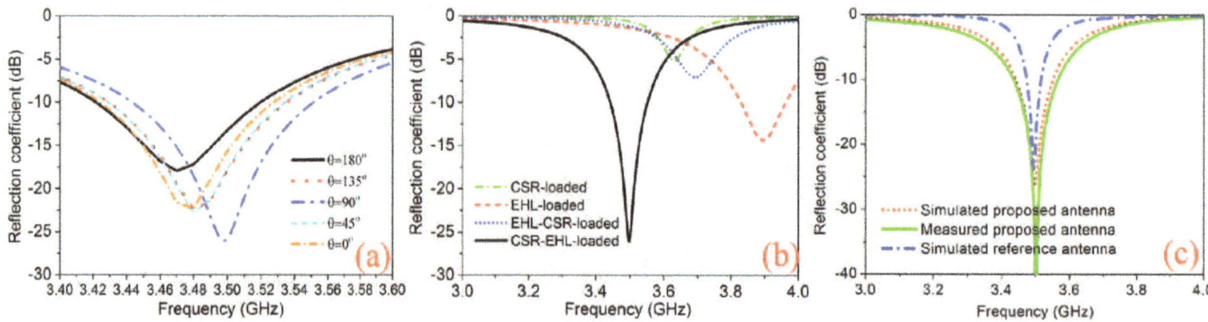

Figure 6. Simulated and measured S-parameters against frequency [6]. The dependence of S-parameters on parameters (a) θ and (b) different kinds of WG-MSs loading. (c) The final simulated and measured results for the reference and proposed antenna.

obtained for the CSR-EHL-loaded (MED-WG-MS) antenna. While for the EHL-CSR-loaded case, by etching the EHL cells in the upper patch structure and CSR elements in the ground plane, a poor impedance matching property at the resonant frequency 3.67 GHz is seen, which can be explained as the asymmetrical property of the MED-WG-MS cell along the *z*-direction. Therefore, we demonstrate that MED-WG-MS elements are the best choice to enhance the antenna performances.

2.4. Numerical and experimental results

With the derived three-step frequency tuning method, we can design an antenna operating exactly at frequency 3.5 GHz as shown in **Figure 4**. For experimental demonstration, we fabricate the antenna with the photograph shown in **Figure 7** [6]. The radiator of the antenna occupies an area of $L_x \times L_y = 20 \times 20$ mm^2, corresponding to $0.20\ \lambda_0 \times 0.20\ \lambda_0$. The size of the ground plane $P \times Q = 40 \times 45$ mm^2. For comparison, the reference antenna is also designed with a patch size of 25.4×27.4 mm^2. The fair comparison indicated that both antennas have an identical size except the radiation patch.

Then, we evaluate the reflection coefficients of the proposed antenna and the reference antenna through the ME7808A vector network analyzer. **Figure 6(c)** illustrates the results of the reflection coefficients. There is a good agreement between the simulation and measurement for the designed antenna. Both the proposed antenna and the reference antenna operate exactly at 3.5 GHz, with the numerical (experimental) resonant dips as -25(-40) dB for the designed antenna and -21.5 dB for the reference one, respectively. The simulated (measured) 10 dB impedance bandwidth is about 115 (132) MHz for the designed antenna, corresponding to 3.29% (3.77%). However, it is 43 MHz for the reference antenna. A simply calculation indicates that the BW has been enhanced significantly by 207% for the designed antenna. **Table 1** shows a detailed comparison between the two antennas, including the patch size, BW, and the values of CF and BIF. Note that the theoretical and simulated CF and BIF are in reasonable agreement with each other. The small CF = 0.57 and large BIF = 2.67 indicate that the designed antenna has the best performance compared with previous magneto-dielectric antennas [1–5]. The higher measured BIF is mainly ascribed to the random errors in measurement. To summarize, the

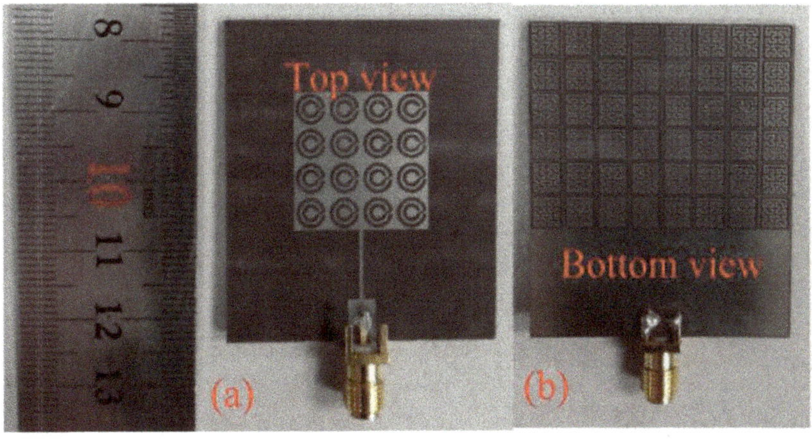

Figure 7. Photograph of the fabricated antenna [6]. (a) Top view and (b) bottom view.

Type	Size	BW	CF	BIF	Antenna efficiency
	(mm²)	(MHz (%))	Theory/simulation	Theory/simulation (measurement)	
Conventional patch antenna	25.4 × 27.4	43 (1.23%)	0.41/0.57	2.87/2.67 (3.07)	95.68%
Our work [6]	20 × 20	115/132 (3.29/3.77%)			93.23%

Table 1. Comparison of the reference and proposed antennas [6].

proposed MED-WG-MS element, coupled with the derived three-step working frequency method provide a guideline to design the antenna with a bandwidth enhancement at any frequency. In addition, the fabrication of the designed antenna is more convenient without using parasitic elements or metallic via holes [13, 14].

There is a link between the slope of the input admittance and the BW for an antenna. A flatter response of the input admittance means a wide impedance-matching property. **Figure 8** shows comparison of the input admittance for the designed antenna and the reference antenna [6]. It is worth noting that both the real part and the imaginary part of the input admittance for the designed antenna show a flatter response than that of the reference antenna, indicating an enhanced BW for the designed antenna.

Next, we evaluate the field distributions for both the designed antenna and the reference one. **Figure 9(a)** and **(b)** shows the comparison of the H-field distributions for both antennas [6]. There is a nearly periodic distribution for the H-field of the proposed antenna, which is attributed to the loading of the MED-WG-MS cells. More importantly, based on the material

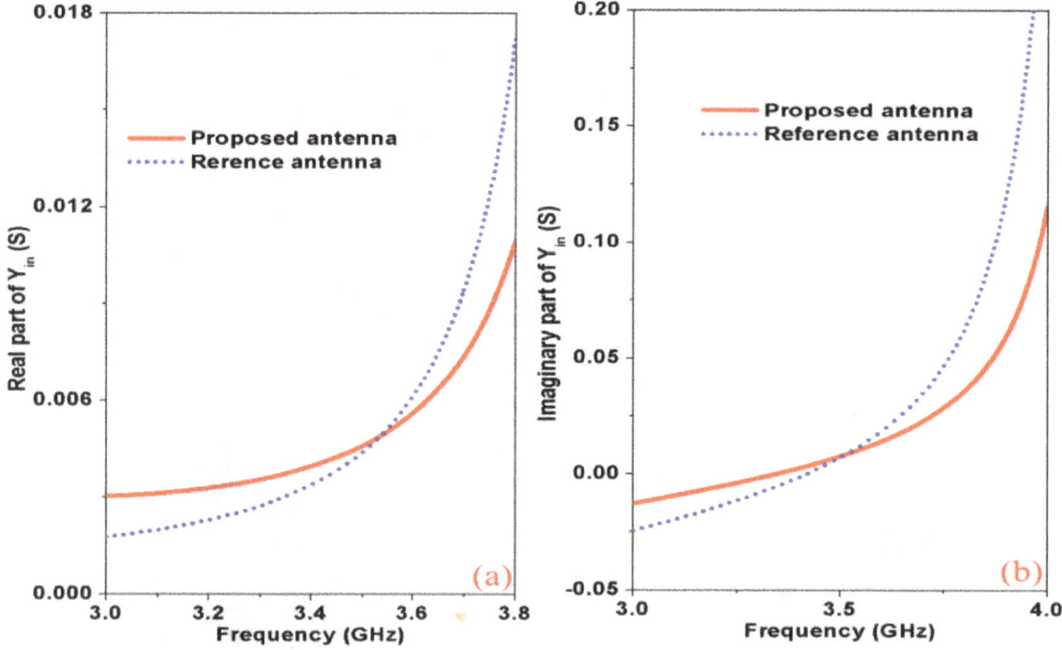

Figure 8. Simulated input admittance for both antennas [6]. (a) Real part and (b) imaginary part.

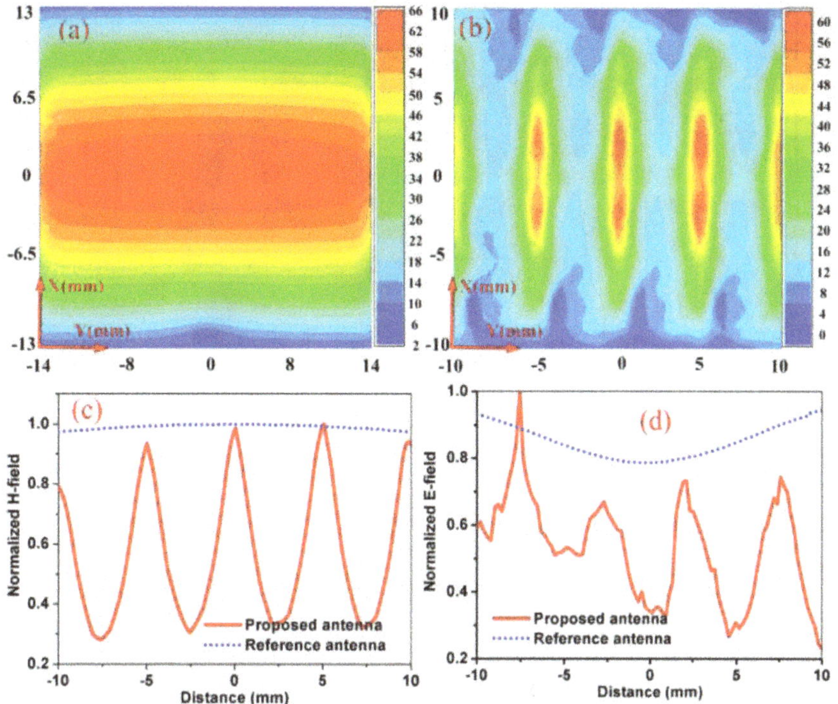

Figure 9. Field distributions for the proposed antenna and reference antenna [6]. The H-field distribution for (a) reference antenna and (b) proposed antenna; a comparison of the normalized (c) E-field intension and (d) H-field intension.

parameters, there is larger wave impedance $\sqrt{\mu_{eff}/\varepsilon_{eff}}$ for the designed antenna. The larger wave impedance induces a weaker H-field intensity and a smaller quality factor, which explained the enhancement of the BW from another aspect. The field intensity along the line $y = 0$ is also studied, as the results shown in **Figure 9(c)** and **(d)**, both the E-field and the H-field intensity are normalized to the maximum value. With the loading of the MED-WG-MS elements, the E-field, and the H-field intensity are periodic for the proposed antenna, while the distributions are almost consistent for the reference antenna. Therefore, we can conclude that the loading of MED-WG-MS elements induces a redistribution of the electromagnetic energy.

Finally, we examine the far-field radiation performances by HFSS simulation and measurement in an anechoic chamber. The three-dimensional (3D) radiation patterns at three representative frequencies (the lower frequency 3.440 GHz, the center frequency 3.5 GHz, and upper frequency 3.555 GHz) are depicted in **Figure 10(a)–(c)** [6]. Given the defects in the ground plane, seemingly bidirectional radiation patterns are observed for the three frequencies. But consistent with the conventional patch antenna, the designed one still works at TM_{10} mode. The radiation patterns in both principal planes at 3.5 GHz are measured, as the results shown in **Figure 10(d)**. The level of the cross-polarization is better than -22.8 dB. Referring to the antenna gain in **Figure 10(e)**, good agreement is observed for the simulated and measured results. Within the 10 dB impedance bandwidth, the gain is higher than 4.6 dBi. Moreover, the radiation efficiency of the antenna is about 93.23%.

In summary, a new strategy of enhancing the BW and reducing the size of a patch antenna is proposed by simultaneously manipulating the material parameters μ_{eff} and ε_{eff}. For proof of

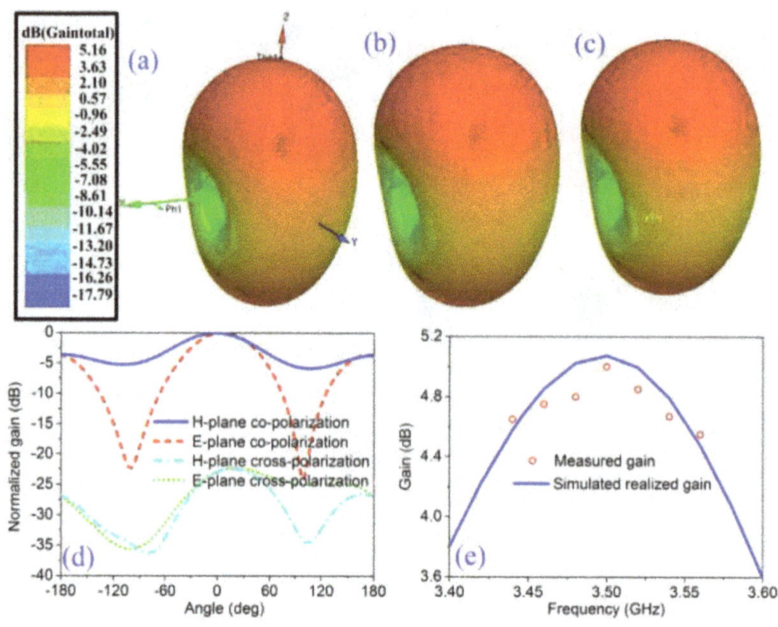

Figure 10. Radiation patterns for the designed antenna [6]. Simulated 3D radiation patterns at (a) 3.440 GHz, (b) 3.50 GHz and (c) 3.555 GHz; (d) measured 2D radiation pattern at 3.50 GHz; (e) simulated and measured antenna gain.

the concept, a MED-WG-MS element is designed to realize larger values of refractive index and wave impedance by tuning the geometrical parameters. Loading these elements in a patch antenna, about 307% BW and 42.53% miniaturization are obtained. This chapter provides a new avenue to design patch antenna with enhanced bandwidth and also compact size.

3. Miniaturized circularly polarized antenna with fractal meta-surface and fractal resonator

We have discussed about the strategy of employing MED-WG-MS element to manipulate the effective material parameters μ_{eff} and ε_{eff} in the last subsection. In this section, we explore a new scheme to realize miniaturized circularly polarized (CP) antenna with fractal meta-surface and fractal resonator. The fractal meta-surface is used to design reactive impedance surface (RIS) to improve the antenna performance in terms of reducing substrate thickness and achieving good front-to-back ratio, whereas fractal resonator with strong space-filling property achieves the CP property and further size reduction.

Unlike linearly polarized (LP) antennas, microstrip circularly polarized (CP) antennas have a stable date transmission rate regardless of the polarizations of the transmitter and the receiver, thus have found numerous applications in the recent wireless communication system [15, 16]. However, traditional CP antennas are designed by involving various perturbations (such as truncated corners), which conflicts with the miniaturization requirement and wide bandwidth applications [17, 18]. In recent years, meta-surfaces, such as reactive impedance surface (RIS), have been applied to enhance the performances of CP antennas [19–28], such as realizing miniaturization by loading RIS [20], achieving multi-frequency operation [21, 25]. However, there are never open reported techniques to reduce the profile of the CP antenna. In this section, fractal concept has been introduced to design RIS, which realizes both miniaturization and low profile. Good performances of the designed CP antenna by using the fractal RIS are numerically and experimentally demonstrated.

First, we show the perspective view of the proposed Hilbert fractal RIS (HRIS)-inspired Wunderlich-shaped fractal complementary split ring resonator (WCSRR)-loaded CP antenna [29], as shown in **Figure 11** [29]. The antenna consists of three layers, the upper metallic radiator, the HRIS spacer, and the lower metallic ground plane. WCSRR is etched in the metallic radiator to realize both antenna miniaturization and CP wave. 6 × 6 HRIS elements are loaded

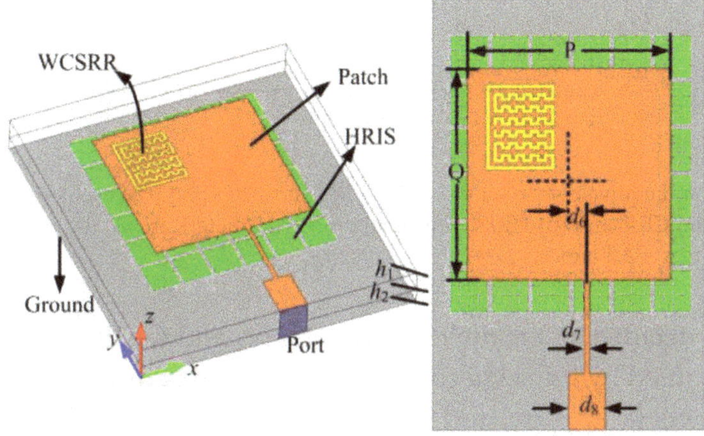

Figure 11. Topology of the proposed CP antenna based on HRIS and WCSRR slot [30]. (a) Perspective; (b) top and bottom view as well as the illustration of geometrical dimensions. The detailed geometrical parameters are listed as: $P = Q = 21.5$ mm, $h_1 = 1.5$ mm, $h_2 = 1$ mm, $d_6 = 2$ mm, $d_7 = 0.6$ mm and $d_8 = 4.1$ mm.

on the F4B substrate under the patch radiator. Two inexpensive F4B dielectric layers ($\varepsilon_r = 2.65$, $\tan\delta = 0.001$) are adopted in the proposed antenna with $h_1 = 1.5$ mm and $h_2 = 1$ mm. The overall antenna occupies a volume of 40 mm × 45 mm × 2.5 mm, and the patch radiator has a dimension of 21.5 mm × 21.5 mm. The working frequency of the antenna is chosen as 3.5 GHz (Wimax band). For characterization, we perform numerical simulations using a finite element method (FEM)-based solver Ansoft HFSS.

3.1. The EM property of the fractal RIS

RIS, proposed by Kamal et al. [22], has been used widely in antennas to reduce the radiator size and enhance the working bandwidth. In **Figure 12(a)** [29], conventional RIS (CRIS) consists of a periodic system of metallic patches ($l_a \times l_b = 4.3$ mm × 3.8 mm), 1-mm-thick F4B spacer, and a metallic ground plane. The upper substrate is introduced to support the patch radiator. In simulation, an infinite CRIS meta-surface is illuminated by transverse electric and magnetic (TEM) plane wave with the propagation vector along the *z*-direction and the electric field along the *y*-direction. Therefore, perfect electric conductor (PEC) conditions are assigned in front and back boundaries, and perfect magnetic conductor (PMC) conditions in left and right boundaries, respectively. Under this circumstance, the CRIS element can be represented by a parallel resonant tank which consists of an inductor *L* and a capacitor *C*. The

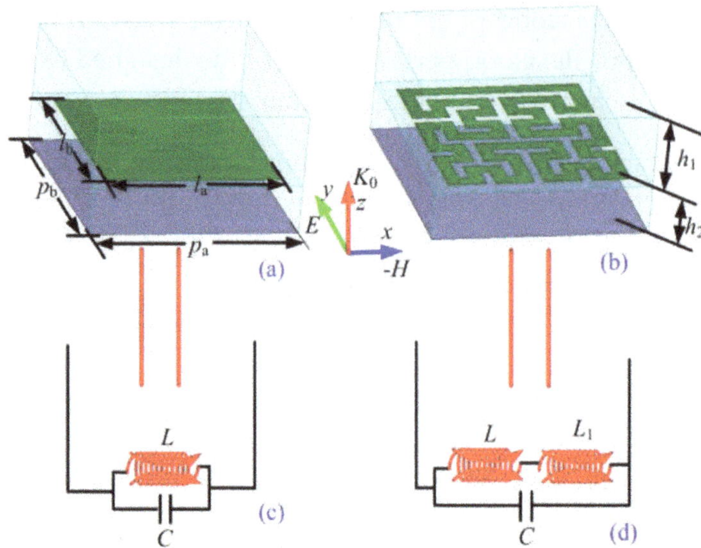

Figure 12. The topologies and equivalent circuit models of CRIS and HRIS [30]. The topologies of (a) CRIS and (b) HRIS; relative circuit model for (c) CRIS and (d) HRIS. The geometrical parameters are list as: h_1 = 1.5 mm, h_2 = 1 mm, l_a = 4.3 mm, l_b = 3.8 mm, p_a = 5 mm, p_b = 4.5 mm.

inductor L is mainly decided by the dielectric constant ε_r and the thickness h_2. While the shunt capacitor C is determined by the edge coupling between the adjacent cells. A variation of the geometrical parameters (l_a, l_b, p_a, p_b, and h_2) induces a change of the lumped circuit elements, and thus the resonant frequency. We can increase the lumped inductor L by introducing compact structures in the periodic patch, corresponding to reduce the substrate thickness h_2. For example, mushroom-like RIS increases the inductor L by digging the metallic via holes which brings about considerable losses for the concentrating currents around the holes [22]. Here, we propose the concept fractal RIS to increase the inductor L. An additional inductor L_1 is introduced to represent the EM response of the fractal structure. In fact, the fractal structure can be chosen arbitrary, e.g., the Sierpinski curve and Koch curve. Here, we adopt the Hilbert RIS, as the topology shown in **Figure 12(b)**. The value of L_1 is determined by the iteration orders (IO). **Figure 13** shows three kinds of HRISs with IO = 0, 1, and 2, respectively. Different from the traditional Hilbert fractal curve, these curves are revised by extending the start and the end points with the length of a minimum fractal segment, and then connect the terminals.

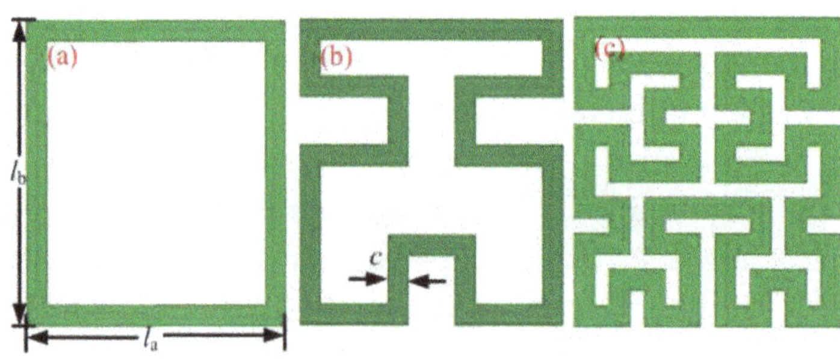

Figure 13. The revised Hilbert curves with different IOs (a) IO = 0; (b) IO = 1; (c) IO = 2 [30]. The geometrical parameters are listed as: l_a = 4.3 mm, l_b = 3.8 mm, c = 0.4 mm.

To evaluate the EM response of the HRIS, **Figure 14(a)** depicts the reflection phases of HRIS with different IOs. Note that the fair comparison is made with the geometrical parameters consistent except the IOs of the HRISs. At the resonant frequency, the reflection phase is about zero [27]. As seen from **Figure 14(a)**, the operating frequency changes from 9.1 to 7.41 GHz as IO increases from 0 to 2. The second HRIS cell occupies a dimension of $\lambda_0/8.1 \times \lambda_0/9 \times \lambda_0/16.2$, which is much smaller than the working wavelength. With a larger IO of the HRIS, stronger space-filling property of the structure is obtained, which induces a smaller resonant frequency. Here, IO = 2 is selected from the tradeoff of an easy fabrication and an electrically smaller structures. **Figure 14(b)** illustrates the dependence of reflection phase for RISs on the substrate thickness h_2. The resonant frequencies decrease as h_2 increases. It is worth noting that the HRIS with h_2 = 1 mm has an almost same resonant frequency with that of the CRIS with h_2 = 2 mm, which means a reduction of substrate thickness for the HRIS. Compared with the conventional RIS, fractal RIS has huge advantages in smaller size, lower thickness, and also easier working frequency control. Since more circuit parameters are inserted by the fractal structure, more freedom is provided to tune the working frequency.

3.2. Working principle of compact fractal resonator

With a unique EM property and a planar structure, split ring resonator (SRR), and its complementary part, CSRR have been utilized in designing microwave devices and enhancing their performances. However, it is still a great challenge to excite CP wave by integrating the fractal strategy with the CSRR structure. We combine the Wunderlich-shaped fractal structure and the CSRR (WCSRR) to achieve not only antenna miniaturization but also good CP radiation. To demonstrate the advantage of WCSRR over the traditional CSRR, the EM response of three kinds of CSRRs is compared. **Figure 15** presents the basic topologies of different SRRs, such as conventional SRR, meander-line-loaded SRR (MSRR), and WSRR. Based on the Babinet principle, we etched the CSRR, MCSRR, and WCSRR structures in the patch radiator. The reflection coefficients and axial ratio (AR) for antennas with different slots are provided in **Figure 16**. Good impedance matching property for all cases can be observed clearly by the

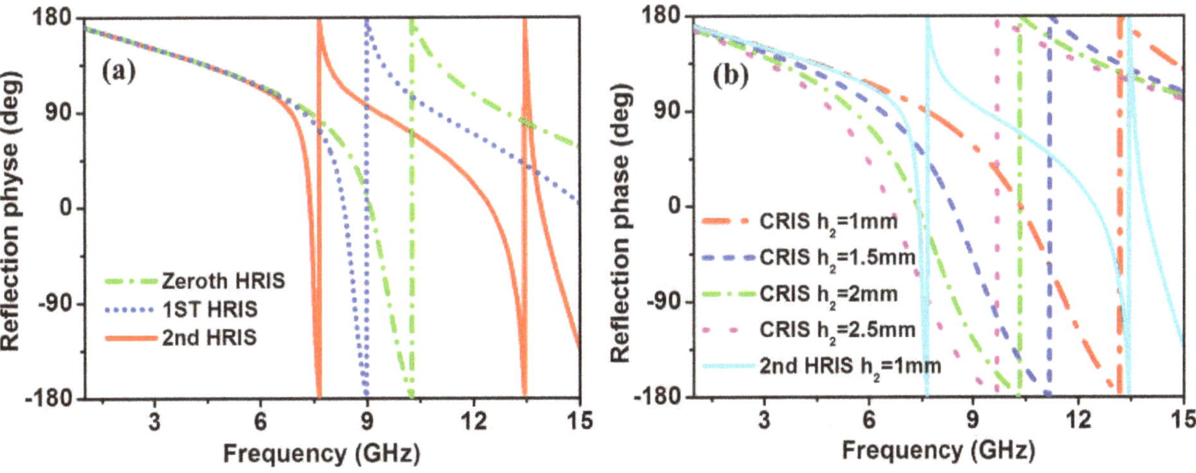

Figure 14. Numerically simulated reflection phase [30]. (a) The dependence of reflection phase on IOs; (b) the dependence of reflection phase on h_2.

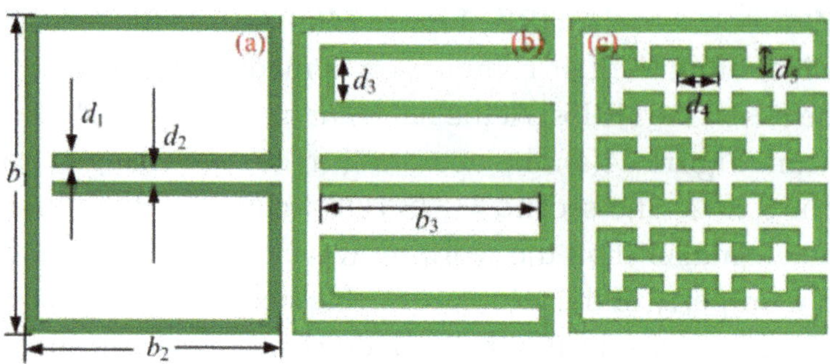

Figure 15. Schematic of the proposed revised SRRs [30]. Topology of (a) SRR, (b) MSRR and (c) WSRR as well as the illustration geometrical dimensions. The final geometrical parameters are listed as: $d_1 = d_2$=0.4 mm, d_3 = 1.1 mm, d_4 = 1.2 mm, d_5 = 0.9 mm, b_1 = 9 mm, b_2 = 7.6 mm, and b_3 = 6.4 mm.

resonant dips, as shown in **Figure 16(a)**. The conventional antenna without CSRR loading operates at about 3.8 GHz, whereas the center frequency reduces significantly to 3.75, 3.65, and 3.5 GHz for the CSRR-, MCSRR-, and WCSRR-loaded antennas, respectively. The extending of the current path caused by the fractal curve induces the considerable reduction of operating frequency, which is equivalent to realize antenna miniaturization working at the same frequency. **Figure 16(b)** shows the effects of CSRRs on the AR performances. Note that the slots have a significant effect on the antenna AR. An obvious linearly polarized antenna is obtained without loading CSRRs. The values of AR reduce significantly by loading CSRRs. The best performance of the CP antenna is observed by loading WCSRR, which achieves a lower operating frequency and a smaller AR value. Therefore, we can use the Wunderlich-shaped fractal slot to improve CP antenna performances.

3.3. Numerical results of a CP antenna

The feeding position as well as the slot position is very important in determining both input impedance and AR value. Antenna performances are simulated by HFSS via tuning the feeding position. The parameter d_6 increases from 0 to 3 mm in steps of 0.5 mm, the

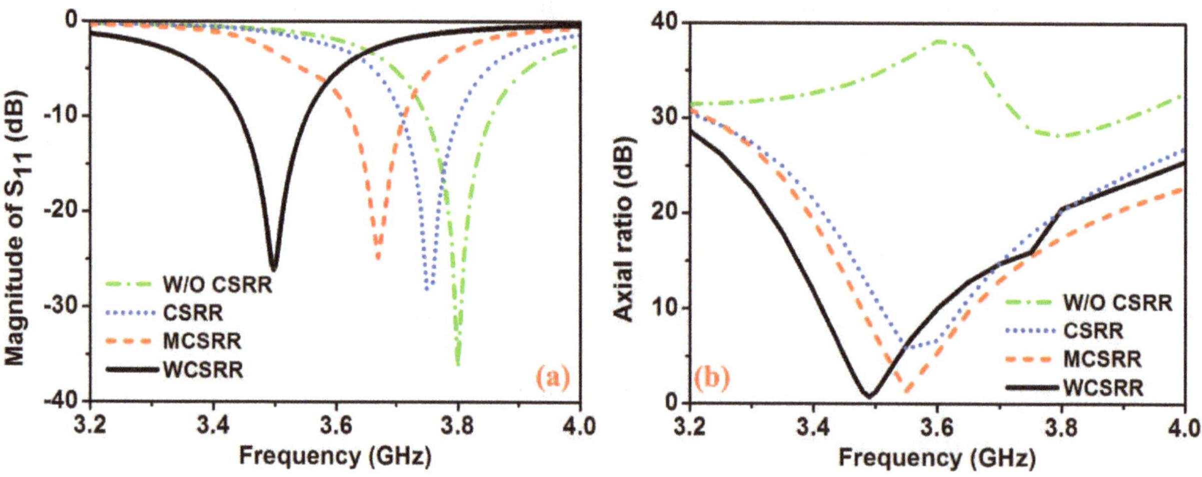

Figure 16. Simulated reflection coefficients and ARs against frequency [30]. (a) The dependence of reflection coefficients on different CSRR structures; (b) the dependence of ARs on different CSRR structures.

corresponding reflection coefficients and AR values are shown in **Figure 17**. The operating frequencies keep almost consistent when d_6 increases from 0 to 2 mm, while the impedance matching property deteriorates slightly. A slight red shift of the working frequency appears as d_6 varies from 2.5 to 3 mm. The AR has been significantly improved as d_6 increases, while the AR valleys decrease continuously from 3.52 to 3.48 GHz. As d_6 = 2 mm, both the dip of the reflection coefficient and the AR valley appear at 3.5 GHz, which is interesting and should be highlighted. **Figure 17(c)** and **(d)** shows the comparison of the magnitude ratio E_y/E_x and phase difference δ_x-δ_y as d_6 varies. Note that E_y/E_x decreases as d_6 increases, so as the δ_x-δ_y. As d_6 = 2 mm, the best performance with a flattest curve of the E_y/E_x and an exact 90° δ_x-δ_y is obtained at frequency of 3.5 GHz. Therefore, d_6 = 2 mm is selected for the final CP antenna design.

Then, we study the electric field distribution of the CP wave at 3.5 GHz, as the results shown in **Figure 18**. The electric field is mainly concentrates on the radiation patch and the WCSRR slot. Left-handed CP wave is observed clearly with a clockwise-rotated electric field. The variation of the field on the WCSRR indicates that the slot plays an essential role in exciting the CP wave. Right-handed CP wave can be excited by changing the position of the slot.

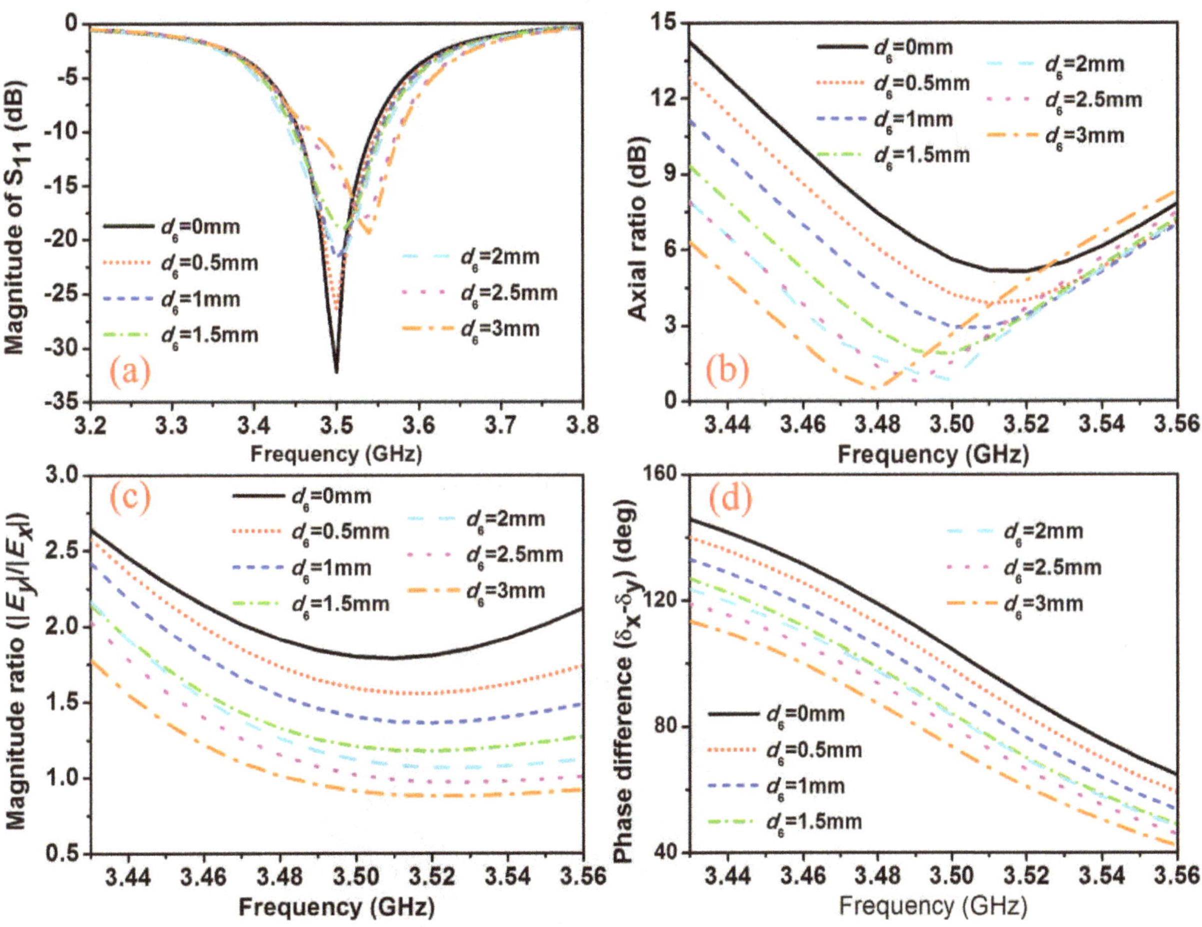

Figure 17. Simulated reflection coefficients, ARs, magnitude ratio as well as phase difference with different feeding positions [30]. The dependence of (a) reflection coefficients, (b) ARs, (c) magnitude ratio, and (d) phase difference on d_6.

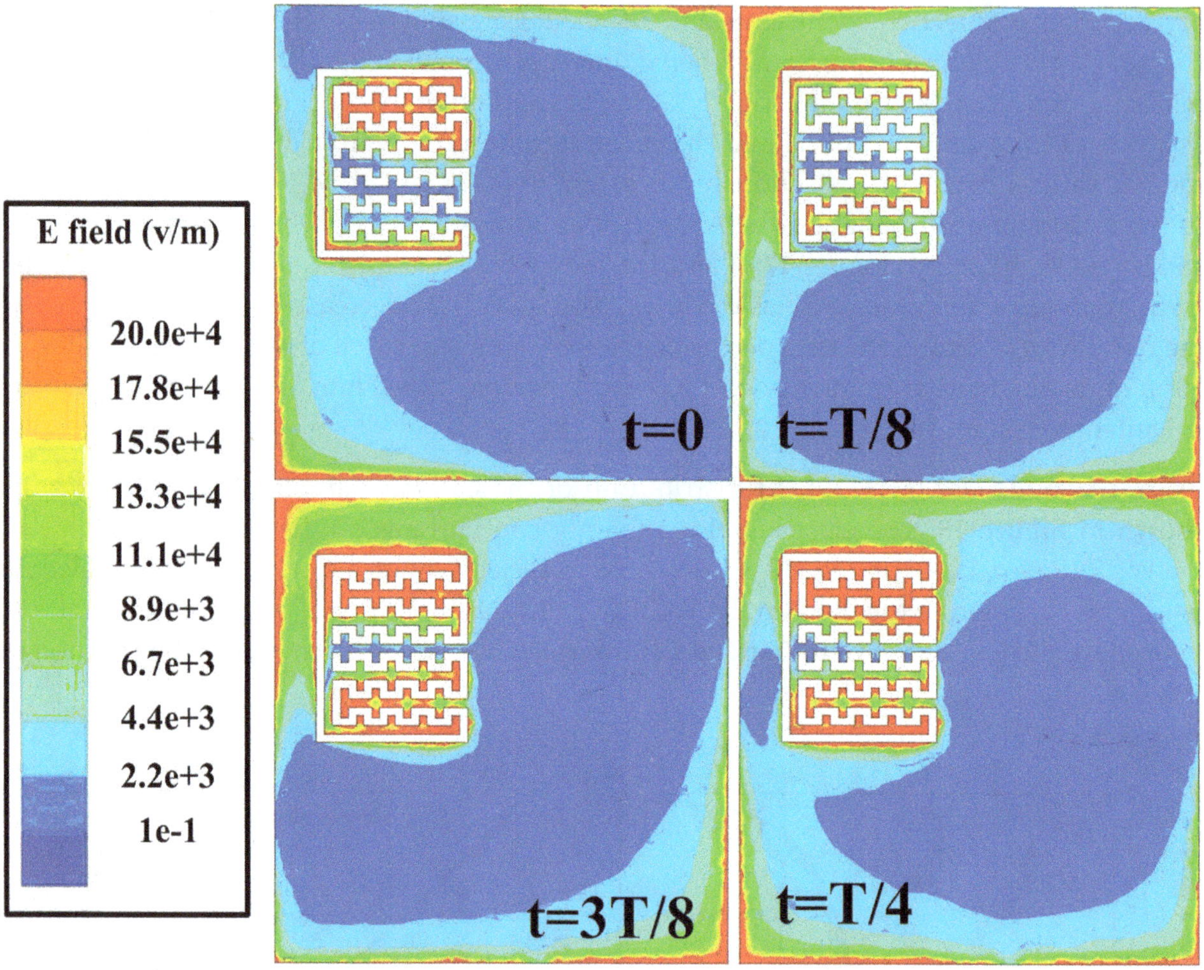

Figure 18. Electric-field distribution in the patch at 3.5 GHz in time-domain [30].

3.4. Fabrication and experimental results

For experimental demonstration, the finally double-layered CP antenna is fabricated based on a standard printed-circuit-board (PCB) technology, with the fabricated sample for the upper layer and the lower layer shown in **Figure 19**. The two layers with the same footprints are boned tightly together by an adhesive. Then we experimentally evaluate the performances of the CP antenna.

Figure 19. Photograph of the fabricated proposed antenna. (a) Top view; (b) bottom view [30].

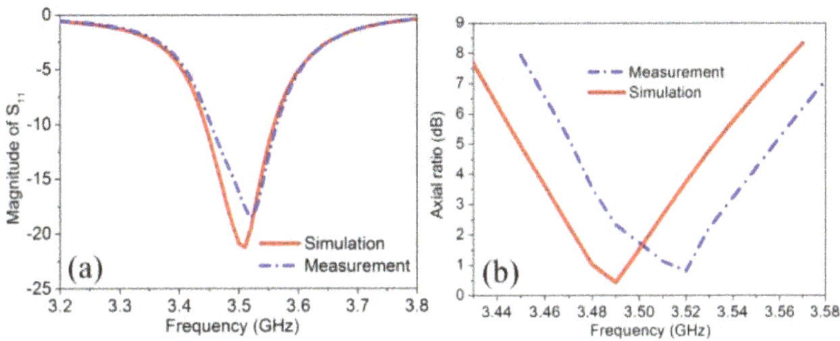

Figure 20. Simulated and measured (a) reflection coefficients and (b) ARs of the proposed CP antenna [30].

First, we examine the impedance matching property of the antenna. The reflection coefficient of the sample is measured by a vector network analyzer (ME7808A). **Figure 20(a)** depicts a comparison of the simulated and measured reflection coefficients. We can see clearly that there is an excellent agreement between the numerical and experimental results. A slight frequency shift upward about 20 MHz in the experiment is mainly attributed to the fabrication errors which is inherent and the introduction of the adhesive in the assembly process. The simulated (measured) reflection coefficient has a resonant dip of -22 dB (-18 dB) at about 3.5 GHz (3.52 GHz), respectively. The working bandwidth, characterized by the 10 dB return loss, is about 132 and 127 MHz for the simulation and measurement, respectively. The relative BW is calculated of 3.77% and 3.61%.

Then, we evaluate the AR ratio of the designed CP antenna. In the experiment, the AR is measured by the intensity ratio between *Ey* and *Ex*, which is obtained in an anechoic chamber through the far-field measurement system. As seen from **Figure 20(b)**, the 3-dB AR bandwidth for the simulated and measured results is 55 and 65 MHz, corresponding to 1.58 and 1.86%, respectively. And the minimum values of the AR are 0.41 and 0.75 for the simulated and measured results. Note that 3-dB AR BW is completely within the 10-dB impedance BW.

Finally, we measure the far-field radiation patterns in *xoz* and *yoz* planes, as the results shown in **Figure 21**. There is a good agreement between the simulation and measured results. The

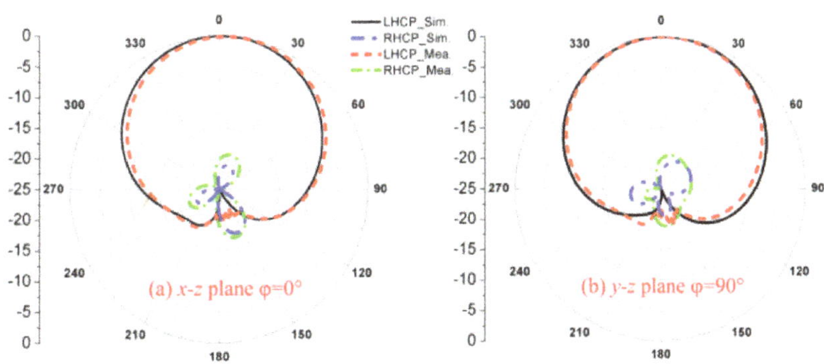

Figure 21. Simulated and measured far-field radiation patterns of the proposed CP antenna at CF in (a) *x-z* plane and (b) *y-z* plane [30].

CP wave is detected as expected. We can see clearly that the direction of the peak radiation is around the boresight direction. A pure CP wave is demonstrated for that the cross-polarization is better than 18.5 dB in both simulated and measured results, which should be highlighted. The front-to-back ratio (F/B) is larger than 15 dB in the experimental result. The radiation gain is about 6.3 dBic for both numerical and experimental results. The designed CP antenna realizes 91.5% antenna efficiency. Good radiation performances, such as good AR property and comparable antenna gain, indicate that the antenna has a potential value in the recent wireless communication system.

In summary, we have proposed a new strategy to design CP antenna by combining a fractal RIS and a fractal slot. For proof of the strategy, a CP antenna by loading HRIS structure and the WCSRR slot is designed, assembled, and measured. The experimental result coincides well with the simulated case. The antenna advances in many aspects such as a compact size (40 mm × 45 mm × 2.5 mm), good AR property (0.75 dB), and comparable radiation gain (6.3 dBic). In addition, the antenna is fabricated based on the PCB technology, free of via holes and without using the complex feeding network.

4. Ultra-thin polarization beam splitter using TGMS

Meta-surfaces, as a 2D planar inhomogeneous metamaterials, are composed of carefully selected elements with specific EM responses, have attracted much attention recently due to their strong abilities to manipulate the wavefront of transmitted and reflected EM waves. Very recently, scientists and engineers have designed functional devices by using meta-surfaces. One of the most important applications is to realize the polarization beam splitter (PBS). PBS is a typical device to manipulate the differently polarized waves independently, which has been found essential in photonics. Conventional PBSs, achieving by natural crystal birefringence [30] and 2D photonic crystals [31, 32], suffer from low efficiencies and limited splitter angle. The PBS performances have been enhanced by using the semiconductor meta-surface [33], photonic-integrated circuits [34], and 2D metamaterials [35–37]. However, these devices are electrically large and complex in fabrication. In this subsection, we propose a novel PBS based on the 2D transmissive phase gradient meta-surface (TPGM) [38]. The proposed PBS is designed based on the generalized Snell's laws, which has a low-profile about 0.1 λ_0, high transmission efficiency, and also convenient wave control.

4.1. Polarization-controlled mechanism of local element

According to the geometrical optics, the EM wave would be reflected or refracted when passing through the interface of two media. When the EM wave propagates in a uniform medium, the wave vector can be written as

$$\begin{cases} k_{xi} = 0 \\ k_{yi} = 0 \\ k_{zi} = k_0 \end{cases} \tag{5}$$

where k_{xi} represents the wave-vector along the x-direction and k_0 is the propagation constant. The generalized laws of reflection and refraction, derived from the Fermat's principle, indicate that the anomalous reflection or refraction phenomenon can be observed at the interface when the phase discontinuity exists [39]. At the same time, the EM response of the GMS is independent when excited with differently polarized waves. Illuminated by a normally incident EM wave along the z-direction with the electric field along the i-direction (x- or y-direction), the wave vector of the transmission can be calculated as

$$\begin{cases} k_{(x,t)}^{(\hat{i})} = \xi_x(i) \\ k_{(y,t)}^{(\hat{i})} = \xi_y(i) \\ k_{(z,t)}^{(\hat{i})} = \sqrt{k_0{}^2 - (k_{(x,r)}^{(\hat{i})})^2 - (k_{(y,r)}^{(\hat{i})})^2} \end{cases} \tag{6}$$

where the superscript i denotes the incident polarization $\vec{E}//\vec{x}$ or $\vec{E}//\vec{y}$ (horizontal and vertical polarizations, $E_{//}$ or $E_{\wedge}$). $\xi_x{}^{(i)}$ represents that the gradient along the x-direction while the incident polarization is along the i-direction.

For the practical GMS, the phase gradient can be calculated as

$$\begin{cases} \xi_x(x) = \dfrac{\partial\varphi_x(x,y)}{\partial x}, & \xi_y(x) = \dfrac{\partial\varphi_x(x,y)}{\partial y} \\ \xi_x(y) = \dfrac{\partial\varphi_y(x,y)}{\partial x}, & \xi_y(y) = \dfrac{\partial\varphi_y(x,y)}{\partial y} \end{cases} \tag{7}$$

According to Eq. (7), two points should be highlighted. First, the phase distribution φ at the position (x, y) can be controlled by both the geometrical structuring and the polarization of the incident EM wave. Second, the phase gradient ξ is determined by the phase distribution $\varphi(x, y)$ at different directions. The phase distribution $\varphi(x, y)$ and the phase gradient ξ provide a powerful freedom to manipulate the EM wave, simultaneously.

Based on Eq. (7), a transmissive PBS can be realized by controlling the parameters $\xi_x(x)$ and $\xi_y(y)$ independently. In other words, we can manipulate the transmission phase $\varphi_x(x, y)$ and $\varphi_y(x, y)$ to achieve desirable gradients ($d\varphi_x/dx$ and $d\varphi_y/dy$) along the x- and y-directions, respectively. According to the generalized Snell's law, the refraction angles to horizontal and vertical polarizations can be calculated as

$$\begin{cases} \theta_{t//} = \arcsin(\dfrac{\lambda_0}{2\pi}\dfrac{d\varphi_x}{dx}) \\ \theta_{t_\perp} = \arcsin(\dfrac{\lambda_0}{2\pi}\dfrac{d\varphi_y}{dy}) \end{cases} \tag{8}$$

To make the principle of the PBS clear, **Figure 22** provides the schematics of anomalous refractive phenomena in four different conditions. 2D TPGMs, positioned at the xoy plane with different phase distributions, are shined by hybrid EM waves along the z-direction, respectively. As seen from **Figure 22(a)**, TPGM1 has no phase gradient along the x- and

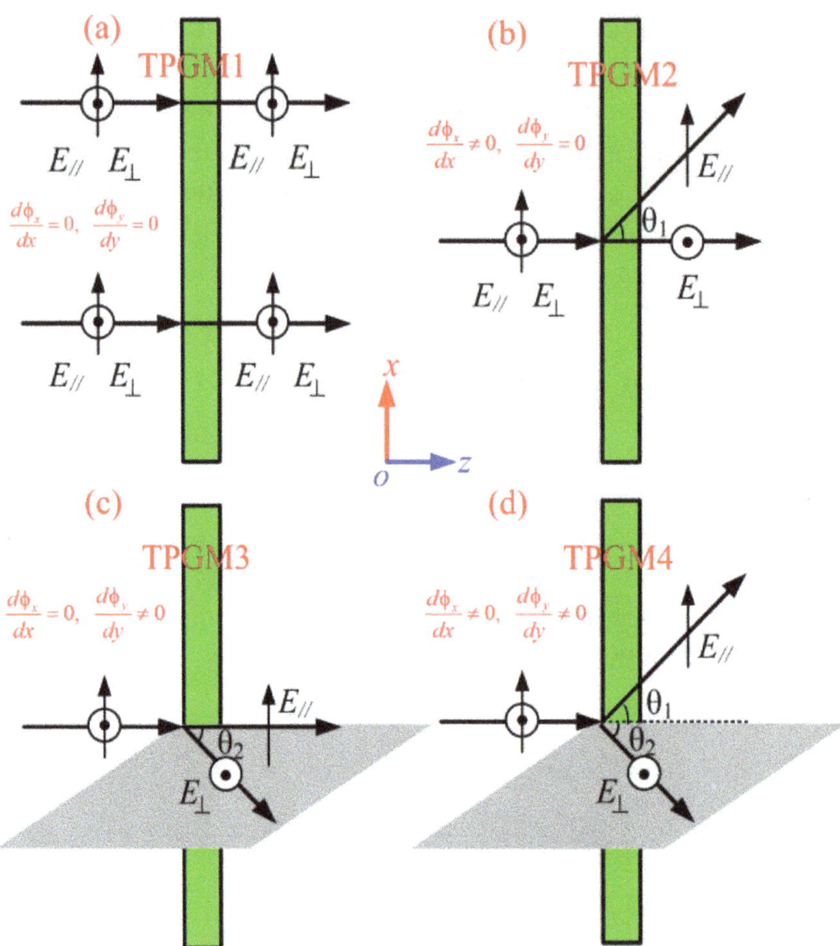

Figure 22. Anomalous refractive effects when 2D TPGMs with different phase distributions are illuminated by the hybrid EM waves [38]. Schematics of the hybrid EM waves passing through (a) the TPGM1 with $d\varphi_x/dx = d\varphi_y/dy = 0$, (b) the TPGM2 with $d\varphi_x/dx \neq 0$ and $d\varphi_y/dy = 0$, (c) the TPGM3 with $d\varphi_x/dx = 0$ and $d\varphi_y/dy \neq 0$, and (d) the PGM4 with $d\varphi_x/dx \neq 0$ and $d\varphi_y/dy \neq 0$.

y-directions ($d\varphi_x/dx = d\varphi_y/dy = 0$). Based on the Fermat's principle, the transmitted wave propagates along the z-direction, and the hybrid EM waves cannot be separated. For the TPGM2 with a phase gradient in the x-direction ($d\varphi_x/ dx\neq 0$) and consistent phase distribution along the y-direction ($d\varphi_y/dy = 0$), the horizontally polarized wave deflected to an angle θ_1, while the vertical polarization wave propagates along the z-direction. The vertical polarization wave can be deflected to angle θ_2 for the TPGM3 with phase gradient existing only along the y-direction ($d\varphi_x/ dx = 0$, $d\varphi_y/dy\neq 0$), as shown in **Figure 22**(c). As expected, hybrid EM waves can be deflected to angles θ_1 and θ_2 when passing through the TPGM4 with phase gradients along the x- and y-directions ($d\varphi_x/ dx \neq 0$, $d\varphi_y/dy \neq 0$). Here, θ_1 and θ_2 are determined by Eq. (8). Therefore, we can realize an arbitrary splitting angle by carefully optimizing the phase distributions on TPGM.

4.2. High-efficiency transparent principle of the transmissive element

For the first step, we should find a transparent element not only with a high transmission coefficient but also a changeable transmission phase. In general, a complete phase shift range

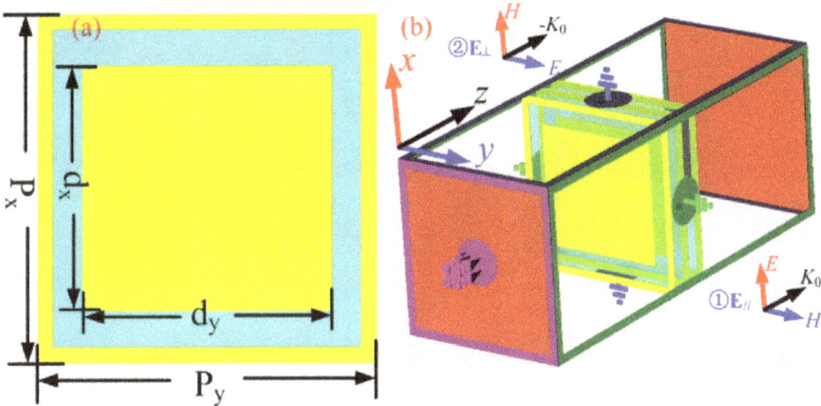

Figure 23. The topology and simulation setup for the proposed tri-player cascaded sub-unit cell [38]. (a) The top view of the subunit cell as well as the geometrical parameters; (b) simulation setup for differently polarized waves.

over 360° is required to guarantee a free wavefront control. Note that EM coupling among the cascaded layers can enlarge the transmission variation range and improve the transmission coefficient [40–44]. Here, three-layer cascaded element is chosen as the basic element, as the topology shown in **Figure 23(a)**. The basic element consists of three identical metallic layers and two intermediate dielectric layers. The commonly F4B substrate is used with a thickness $h = 1.5$mm, dielectric constant $\varepsilon_r = 2.65$, and the loss tangent tan$\delta = 0.001$. We note that although the metallic mesh is not an absolutely necessary element in designing our basic element, its presence can significantly reduce the mutual couplings between adjacent subunit cells, which make our design robust and reliable. The cell has a lattice of $Px \times Py = 11$ mm × 11 mm. We can tune the length d_x and d_y of the inner patch to manipulate the transmission spectra along the x- and y-directions, respectively. We adopt the commercial FDTD solver CST Microwave Studio to obtain the EM property of our element. In simulation, the perfect electric conductor (PEC) boundary conditions are assigned along the x- or y-directions for the horizontally polarized wave or vertically polarized wave, respectively. The absorbing boundary condition is applied along the z-direction, as the simulation setup shown in **Figure 23(b)**.

To obtain the transmission spectra of the basic element, we chose a typical element with $d_x = d_y$ = 8.28 mm. Illuminated by a horizontally polarized plane wave, the transmission magnitude and phase of the element are presented in **Figure 24(a)**. There exists a relatively wide transparency window between 5.4 and 10.35 GHz, and the transmission phase changes smoothly from -131 to -623° within the transparency window. To realize beam deflection for differently polarized waves, different phase gradients along the x- and y-directions should be satisfied. Here, the phase gradients along the x- and y-directions are arranged as $\xi = 0.45\,k_0$ and $\xi = -0.45\,k_0$, respectively, corresponding to 60° phase increment at the x-axis and a 60° phase reduction at the y-axis. By carefully tuning d_x and d_y, the spectra of the chosen elements are shown in **Figure 24(b)**. The transmission coefficients for each chosen element are higher than 0.8, which guarantees a high efficiency of the elements. The polarization-independent property of the element is essential in determining the working efficiency of the device. We verify the polarization-independent property by scanning the parameter d_x from 3 to 8 mm in steps of 1 mm under excitation of different polarizations. As seen from **Figure 25(a)**, the range of about 330° phase variation is obtained as d_x changes for the horizontal polarization, while it is

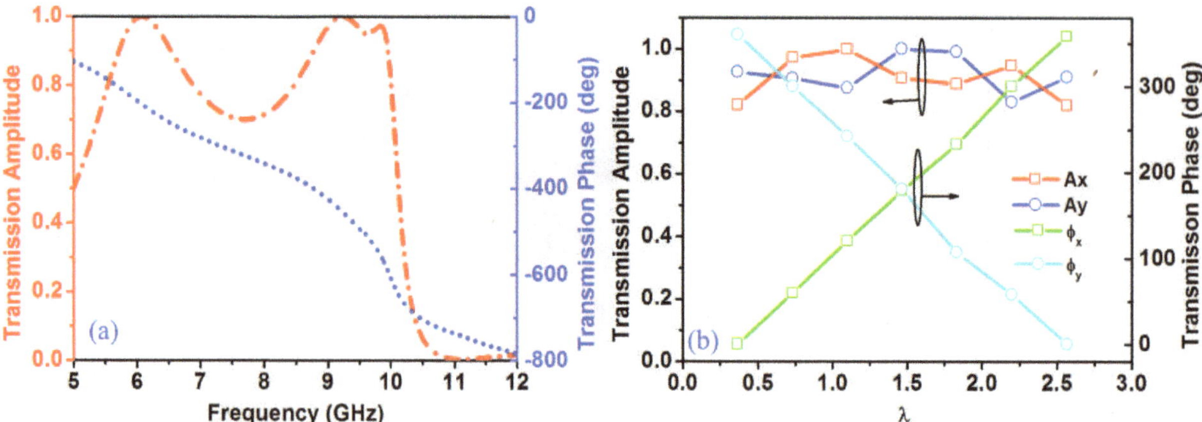

Figure 24. FDTD calculated transmission amplitude and phase for the proposed tri-layer sub-unit cell [38]. (a) Transmission amplitude and phase for the element with $P_x = P_y = 11$ mm, $d_x = d_y = 8.28$ mm; (b) transmission spectra for the elements with 60° phase increment along the *x*-direction and 60° phase reduction along the *y*-direction.

about 20° for the vertically polarized transmitted wave. In addition, the transmission amplitude remains better than 0.8 for all cases. We can conclude that the designed element is polarization-independent based on the phase responses to different polarizations with changeable d_x. The good performance of the element provides us possibilities to design PBS with high efficiency. More importantly, with an independent control of different polarizations, bifunctional meta-surface can be designed with very high performances, which shown great advantages over the conventional meta-surfaces.

4.3. Design and analysis of 2D TGMS

With a well-designed element in hand, we can design the TPGM by carefully chosen the elements satisfying the phases both for horizontal and vertical polarizations. A systematical study on the TPGM can provide a guideline for the upcoming PBS design. We derive a four-

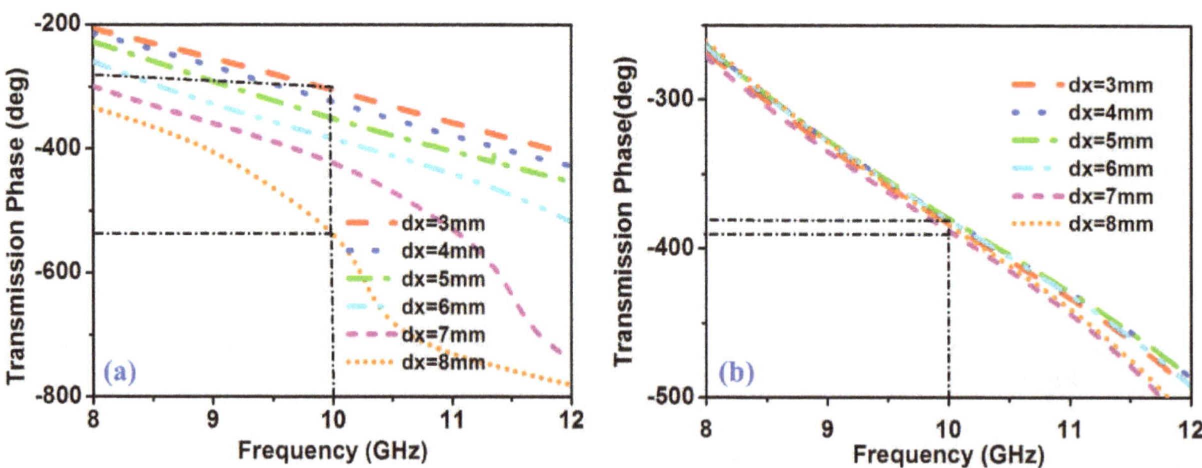

Figure 25. The spectra of transmission phase as functions of frequency and the parameter d_x under excitation of differently-polarized incident waves [38]. The transmission phase response under (a) horizontally polarized and (b) vertically polarized waves.

step design procedure for the general TPGM. For the first step, a transmissive element with a high transmission coefficient and changeable transmission phase should be obtained. Second, the phase distributions at the position (x, y) should be calculated based on the achieving functionalities. Then, the proper subunit cells should be chosen according to the phase distributions. Finally, the EM response of the TPGM should be evaluated by the EM simulator.

Based on derived design procedure, four kinds of TPGM are designed based on the phase distributions, as shown in **Figure 22. Figure 26** provides the electric fields distributions of $E_{//}$ and $E_{\wedge}$ at the working frequency of 10 GHz under excitation of a hybrid EM waves. As shown in **Figure 26(a)**,a conventional meta-surface (TPGM1) without phase gradient is launched by a horn antenna. The transmitted wave of both $E_{//}$ and $E_{\wedge}$ propagate along the z-axis, as shown in **Figure 26(b)** and **(c)**. We can conclude that the TPGM1 is impossible to separate the hybrid polarized waves. As $d\varphi_x/dx$ = -60° and $d\varphi_y/dy$ = 0 for the TPGM2, horizontally polarized wave is deflected to the angle with θ_1 = -27°, which coincides well with the theoretical value obtained by Eq. (8). However, vertically polarized wave is not deflected. Similarly, horizontally polarized wave propagates along the z-direction while vertical polarization is deflected to the y-direction with θ_2 = 27.1° for the TPGM3 with the phase distributions $d\varphi_y/dy$ = 60° and $d\varphi_x/dx$ = 0. For the TPGM4 shown in **Figure 26(j)** with phase gradients along the x- and y-directions ($d\varphi_y/dy$ = 60° and $d\varphi_x/dx$ = -60°), both deflections of θ_1 = -27° and θ_2 = 27° are observed by illuminating with differently polarized waves.

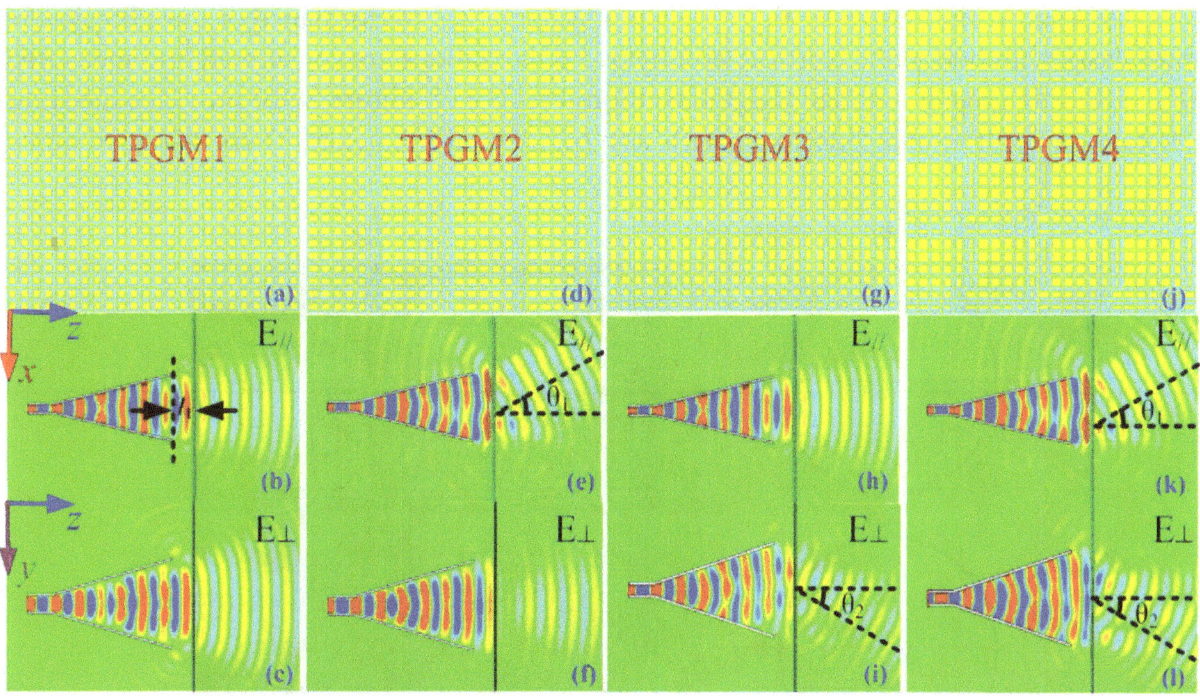

Figure 26. The well-designed four kinds of TPGMs and the results of near-field distributions [38]. The schematics of (a) TPGM1 without phase gradient ($d\varphi_x/dx = d\varphi_y/dy = 0$), (d) TPGM2 with $d\varphi_x/dx = -60°$ and $d\varphi_y/dy = 0$, (g) TPGM3 with $d\varphi_x/dx = 0$ and $d\varphi_y/dy = 60°$ and (j) TPGM4 with $d\varphi_x/dx = -60°$ and $d\varphi_y/dy = 60°$; the electric distributions $E_{//}$ for (b) TPGM1, (e) TPGM2, (h) TPGM3 and (k) TPGM4; the near electric distributions $E_{\wedge}$ for (c) TPGM1, (f) TPGM2, (i) TPGM3 and (l) TPGM4.

4.4. Fabrication and evaluation of PBS

With a well-designed TPGM4, we can further optimize its performance and build a novel PBS by launching the meta-surface by a horn antenna working at the X band. **Figure 27** shows the photograph of the fabricated sample. We can see that 24 × 24 unit cells are adopted for the final PBS, which occupies a total volume of 256 × 256 × 3 mm^3, corresponding to 6.6 λ_0 × 6.6 λ_0 × 0.1 λ_0. It is worth noting that the thickness of the PBS is much smaller than the reported cases. Moreover, the PBS has no complex structures and is fabricated based on the simple print circuit board (PCB) technology. The PBS consists of 4 × 4 super unit cells, as the top and side views of the super unit cell are shown in **Figure 27(a)** and **(b)**. Compared with the working mechanism of the previous PBSs, the designed one is intuitionistic and simple. We can control the differently polarized waves independently due to the different phase gradients at *x*- and *y*-axis, respectively. More importantly, the polarization-independent property of the designed element guarantees the low coupling of different polarizations, which induces a high polarization splitting ratio.

The impedance matching property of the PBS plays an essential role in determining the working efficiency. We evaluate the reflection coefficients of the PBS by launching with different polarizations. First, we assemble the fabricated sample. We used a 20-mm-thick foam plate to support the TPGM and ensure the length l_1 = 20 mm between the TPGM and the horn antenna. The foam has a permittivity near 1 which will not affect the performances of the PBS system. The reflection coefficients are measured by an ME7808A vector network analyzer, and the polarizations of the horn antenna are controlled by the angle θ. The measurement process is shown in **Figure 28(a)**. To obtain a systematical performance of the PBS system, three cases with horizontal, vertical, and hybrid polarizations are considered and measured. Under illuminating of a horizontal polarization with θ = 0°, **Figure 28(b)** provides the simulated and measured reflection coefficients as a function of frequency. There is a good agree-

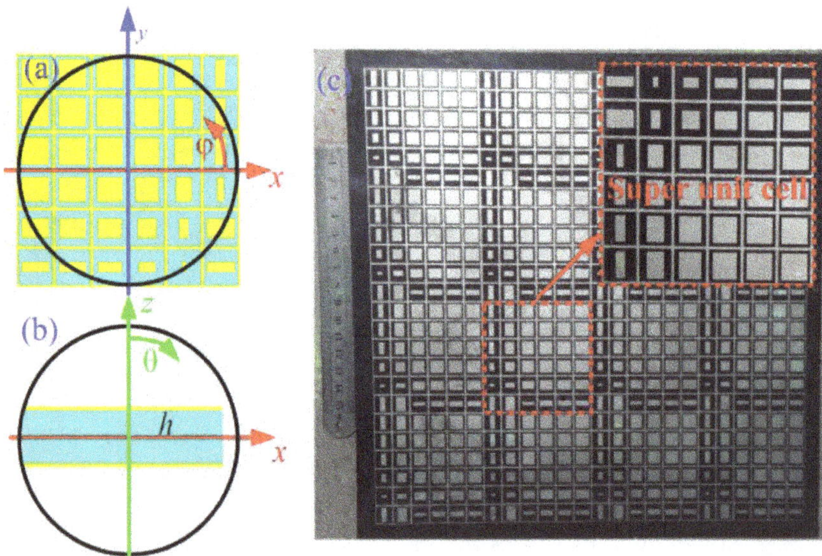

Figure 27. The topology of the unit cell and photograph of the fabricated sample [38]. (a) The top view and (b) the side view of the super unit cell; (c) the photograph of the fabricated sample.

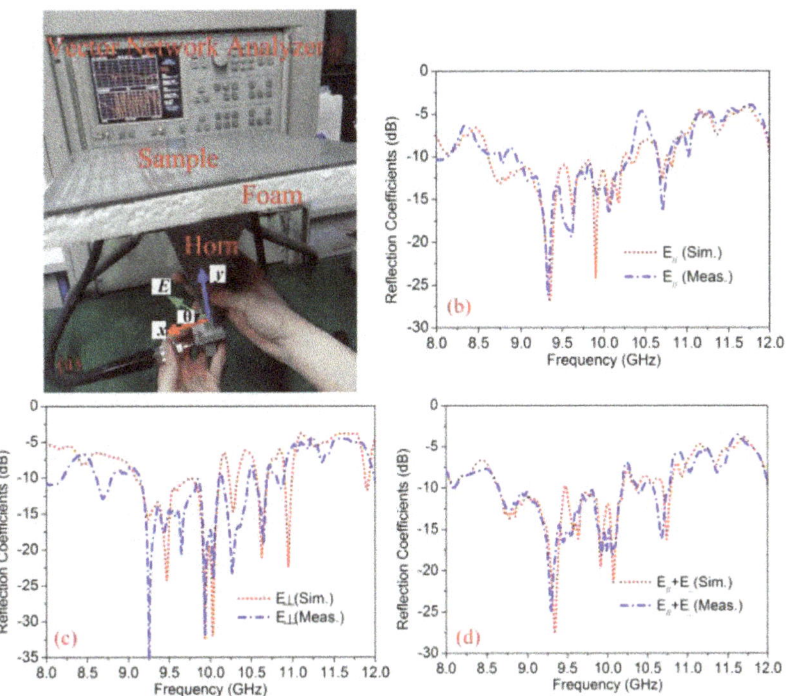

Figure 28. The measurement process of the designed PBS and simulated and measured reflection coefficients for differently polarized incident waves [38]. (a) The measurement process of the designed PBS; the simulated and measured reflection coefficients under illumination of the (b) horizontal, (c) vertical, and (d) hybrid polarizations.

ment between the numerical and measured results. The working bandwidth, ordered by the 10 dB return loss, is about 1.57 GHz (from 8.76 to 10.33 GHz) for the simulation, whereas it is 1.77 GHz (from 8.62 to 10.39 GHz) in measurement. The bandwidth of the PBS is affected by the intersection of the transparency windows of the selected elements. As can be seen from **Figure 28(c)**, the PBS is illuminated by the vertically polarized plane wave with $\theta = 90°$. The slight nonuniformity between the numerical result and measured result is mainly attributed to the fabrication errors. The impedance bandwidth varies from 8.62 to 10.22 GHz and 8.6 to 10.22 GHz for the simulated and measured results, respectively, with the relative bandwidth corresponding to 16.99 and 17.20%. Finally, we rotate the horn antenna by $\theta = 45°$ to obtain a hybrid EM wave. The results are shown in **Figure 28(d)**. The 10-dB impedance bandwidth can be observed clearly as 18.81 and 18.66% for the simulated and measured results, respectively. We can conclude that the designed PBS is insensitive to the polarization of the incident waves, and the PBS has a relatively wide bandwidth.

Then, we examine the 3D far-field patterns of the novel PBS at its working frequency of 10 GHz under excitation of differently polarized waves. Without the TGMS, the horn antenna radiates a narrow beam along the *z*-direction, which can be demonstrated in **Figure 29(a)**. Shining the TGMS with an *x*-polarized incident wave, the transmitted beam is deflected to an angle with $\theta_1 = -26.7°$, which is consistent with the theoretical prediction, as shown in **Figure 29(b)**. One point should be highlighted that the side lobe level has not been deteriorated due to the precise design of the PBS. Shining the TGMS with a *y*-polarized incident wave, the radiation beam propagates along $\theta_2 = 27°$, which shows good agreement with the theoretical value based on Eq. (8) for a second time. Finally, the hybrid polarizations are used

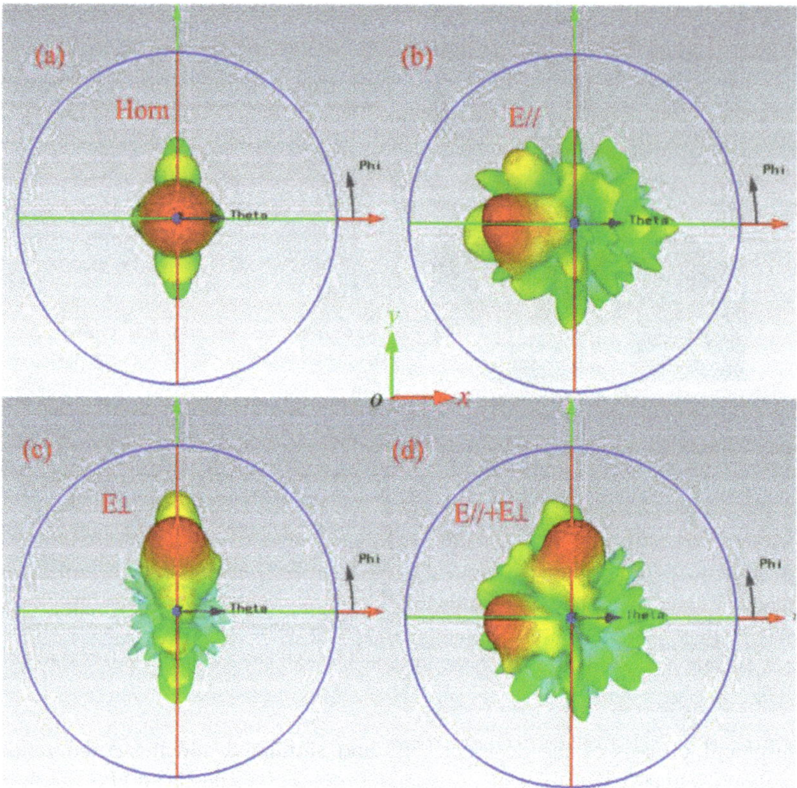

Figure 29. The simulated 3D far-field radiation pattern operating at 10 GHz [38]. (a) The radiation pattern of the bare horn antenna; the 3D patterns of the PBS excited by the (b) horizontally polarized, (c) vertically polarized, and (d) the hybrid polarized waves.

to excite the PBS. **Figure 29(d)** shows two separated beams along the *x*- and *y*-directions. Here, two points should be highlighted. First, the two deflected beams have consistent radiation angles with the theoretical values, which demonstrate the reasonable of the PBS design. Second, the radiation dip between the two main lobes is 30 dB lower than that of the radiation peaks, which indicates an excellent polarization splitting performance. To summarize, the well-designed PBC is able to split the orthogonally polarized wave successfully.

Next, we evaluate the 2D radiation patterns of the novel PBS through the far-field measurement system in an anechoic chamber. Here, three excitation polarizations are considered to investigate the far field performances of the designed PBS, as the simulated and measured results shown in **Figure 30**. The measured results coincide well with the simulation except the radiation pattern in the *xoy* plane as shown in **Figure 30(c)**. The measured beam radiates along θ_1 = -30°, which is slightly departed from the simulated result with θ_1 = -27°. The difference comes from the inherent errors in measurement. In all simulated cases, the levels of the sidelobe are at least 12 dB lower than that of the mainlobe, while it is better than 10 dB for the measured case. Moreover, the front-to-back ratio is about 16 (13) dB for the simulation (measurement). The cross-polarization levels for all cases remain less than 15 dB. The good performances of the designed PBS provide potential applications in wireless communication systems.

The polarization splitting ratio, defined as the radiation gain between two separated radiation beams, is a very important factor for the PBS. We simulate and measure the radiation patterns at the conical surface θ = -27° to evaluate the polarization splitting ratio. As shown in **Figure 31**, the dip between two radiation peaks is about 30 (18) dB for the numerical (experimental) result, which can evaluate the beam separation degree clearly. Finally, we measure the working bandwidth of the designed PBS by examining the radiation patterns under different polarizations, as the results shown in **Figure 32**. We can see obviously that similar

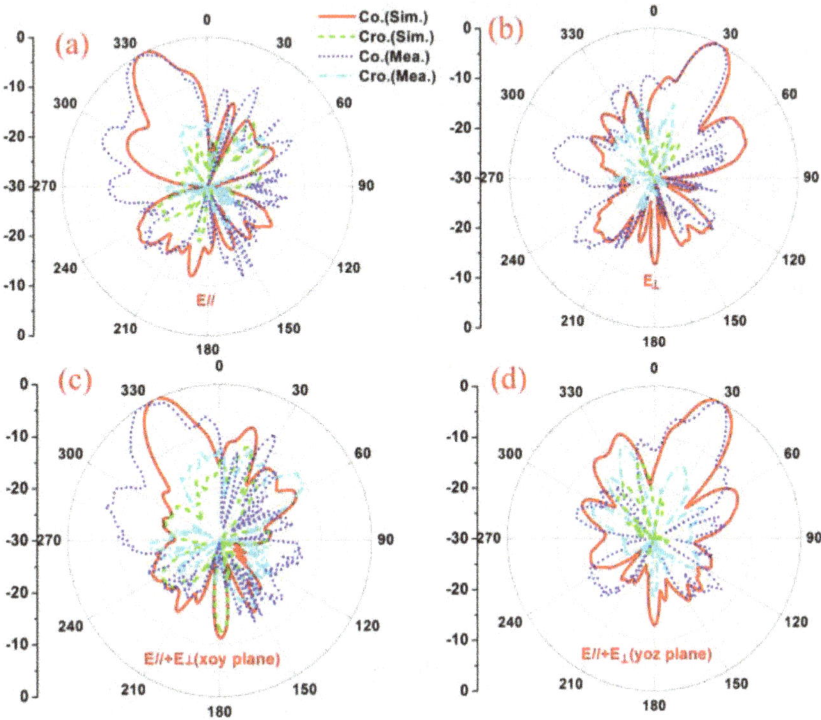

Figure 30. The simulated and measured 2D radiation patterns operating at 10 GHz [38]. The 2D radiation patterns of the PBS excited by the (a) horizontally polarized wave, (b) vertically polarized wave, and (c, d) the hybrid polarized wave.

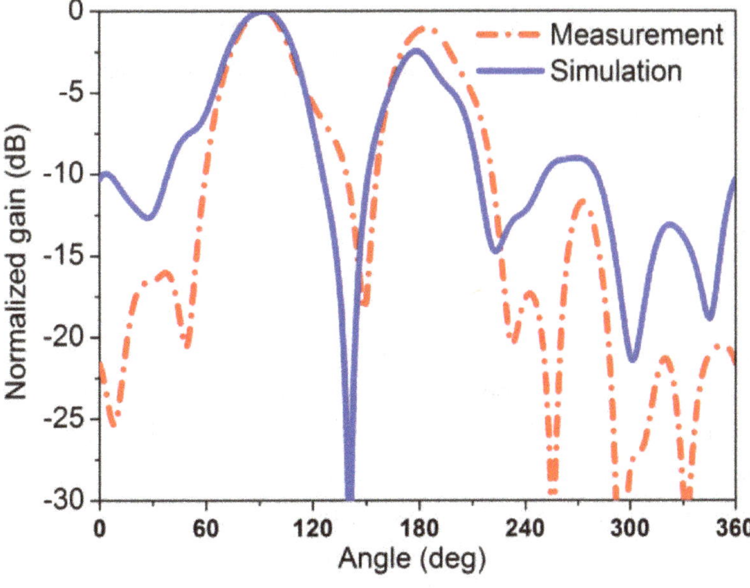

Figure 31. The simulated and measured polarization separation ratios operating at 10 GHz [38].

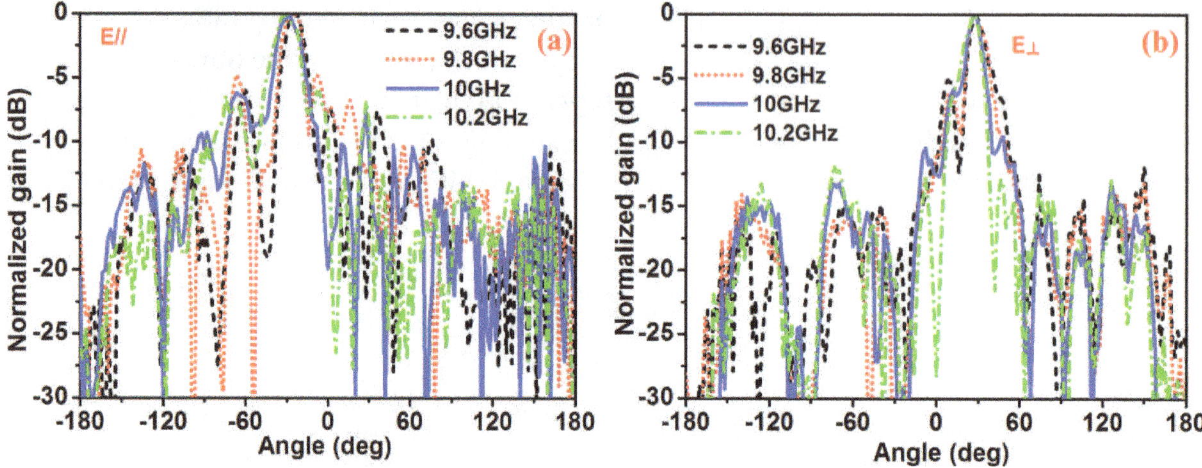

Figure 32. The measured 2D far-field radiation patterns at frequencies changing from 9.6 to 10.2 GHz [38]. The radiation patterns under illumination of (a) TE polarized wave and (b) TM polarized wave.

radiation patterns are detected between 9.6 and 10.2 GHz for both horizontal polarization and vertical polarization, respectively. The half-power beam width (HPBW) is evaluated as about 16° in all cases. The radiation direction of the measured deflected beams varies -26.5° at 9.6 GHz to -28° at 10.2 GHz for the horizontal polarization, while it changes from 27° at 9.6 GHz to 29° at 10.2 GHz for the vertical polarization. The working bandwidth is obtained of about 600 MHz, corresponding to 6%.

In summary, we propose a new strategy to design the PBS aiming at reducing the structure thickness, realizing good polarization splitting ratio, and also high efficiency. A TPGM is proposed based on a carefully designed element, which realizes a high transmission coefficient of more than 0.8, a wide phase variation range over 330° and also a polarization-independent property. The TPGM achieves good beam separation performances as excited by different polarizations. Launched by a wideband horn antenna with a suitable length, an ultra-thin PBS is designed, fabricated, assembled, and measured. Numerical and experimental results show that the PBS can deflect incident waves with different polarizations to different directions with high polarization splitting ratio.

5. Conclusions

In this chapter, we have reviewed our recent efforts in utilizing electrically small meta-surface elements to improve antenna performances and design functional devices, among which three most important aspects have been investigated in depth. First, the constitutive material parameters are controlled by the proposed MED-WG-MS elements, achieving a compact microstrip antenna with enhanced bandwidth. Second, fractal meta-surface and fractal resonator are combined to achieve a CP antenna with a low profile. Third, a high-performance PBS has been proposed based on the TGMS, which shows advances in many aspects such as separating and controlling the orthogonally polarized waves with a polarized splitting ratio better than 18 dB, obtaining a comparable bandwidth of more than 600 MHz, and also gaining high transmission efficiency. Our results pave a new avenue for both engineers and scientists to realize their devices or demonstrate their findings.

Acknowledgements

This work was supported by the National Natural Science Foundation China under Grant Nos. 61372034, and 61501499 and also the Natural Science Foundation of Shaanxi Province under Grant Nos. 2016JM6063 and 2016JQ6001.

Author details

Tong Cai*, He-Xiu Xu*, Guang-Ming Wang and Jian-Gang Liang

*Address all correspondence to: caitong326@sina.cn and hxxuellen@gmail.com

Microwave Laboratory, Air Force Engineering University, Xi'an, China

References

[1] PM. T. Ikonen, S. I. Maslovski, S. A. Tretyakov, et al. "On artificial magneto-dielectric loading for improving the impedance bandwidth properties of microstrip antennas," *IEEE Trans. Antennas Propag.*, Vol. 54, No. 6, pp. 1654–1662, 2006.

[2] X. M. Yang, Q. H. Sun, T. J. Cui, et al. "Increasing the bandwidth of microstrip patch antenna by loading compact artificial magneto-dielectrics," *IEEE Trans. Antennas Propag.*, Vol. 59, No. 2, pp. 373–378, 2011.

[3] H. Mosallaei, K. Sarabandi. "Design and modeling of patch antenna printed on magneto-dielectric embedded-circuit metasubstrate," *IEEE Trans. Antennas Propag.*, Vol. 55, pp. 45–52, 2007.

[4] H. Mosallaei, K. Sarabandi. "Magneto-dielectrics in electromagnetics: concept and applications," *IEEE Trans. Antennas Propag.*, Vol. 52, No. 6, pp. 1558–1567, 2004.

[5] X. M. Yang, X. G. Liu, X. Y. Zhou, T. J. Cui, "Reduction of mutual coupling between closely packed patch antennas using waveguided metamaterials," *IEEE Antennas Wireless Propag. Lett.*, Vol. 11, pp.389–392, 2012.

[6] T. Cai, G.-M. Wang, F.-X. Zhang, et al. "Compact microstrip antenna with enhanced bandwidth by loading magneto-electro-dielectric planar waveguided metamaterials," *IEEE Trans. Antennas Propag.*, Vol. 63, No. 5, pp. 2306–2311, 2015.

[7] C. A. Balanis. Antenna theory: analysis and design, 2nd ed., New York: Wiley, 1997, ch. 14, pp. 36–752.

[8] T. Cai, G.-M. Wang, J.-G. Liang. "Analysis and design of novel 2D transmission line metamaterial and its application to compact dualband antenna," *IEEE Antennas Wireless Propag. Lett.*, Vol. 13, pp. 555–558, 2014.

[9] H.-X. Xu, G.-M. Wang, M.-Q. Qi, L.-M. Li, T.-J. Cui, "Three-dimensional super lens composed of fractal left-handed materials," *Adv. Opt. Mater.*, Vol. 1, pp. 495–502, 2013.

[10] D. R. Smith, S. Schultz, P. Markos, C. M. Soukoulis, "Determination of effective permittivity and permeability of metamaterials from reflection and transmission coefficients," *Phys. Rev. B*, Vol. 78, No. 12, pp. 121102, 2008.

[11] X. D. Chen, T. M. Grzegorczyk, B. I. Wu, et al. "Robust method to retrieve the constitutive effective parameters of metamaterials," *Phys. Rev. E*, Vol. 70, pp.016608, 2004.

[12] H.-X. Xu, G. -M. Wang, Q. Liu, J.-F. Wang, J.-Q. Gong, "A metamaterial with multi-band left handed characteristic," *Appl. Phys. A*, Vol. 107, No. 2, pp. 261–268, 2012.

[13] M. A. Antoniades, G. V. Eleftheriades, "A folded-monopole model for electrically small NRI-TL metamaterial antennas," *IEEE Antennas Wireless Propag. Lett.*, Vol. 7, pp.425–428, 2008.

[14] L.-W. Li, Y.-N. Li, T.-S. Yeo, et al. "A broadband and high-gain metamaterial microstrip antenna," *Appl. Phys. Lett.*, Vol. 96, pp. 164101, 2010.

[15] B.-C. Park, J.-H. Lee. "Omnidirectional circularly polarized antenna utilizing zeroth-order resonance of epsilon negative transmission line," *IEEE Trans. Antennas Propag.*, Vol. 59, No. 7, 2717–2720, 2011.

[16] R. L. Li, J. Laskar, M. M. Tentzeris, "Broadband circularly polarized rectangular loop antenna with impedance matching," *IEEE Microw. Wireless Compon. Lett.*, Vol.16, No. 1, pp. 52–54, 2006.

[17] S.-T. Ko, B.-C. Park, J.-H. Lee, "Dual-band circularly polarized patch antenna with first positive and negative modes," *IEEE Antennas Wireless Propag. Lett.*, Vol. 12, pp.1165–1168, 2013.

[18] K.-P. Yang, K.-L. Wong, "Dual-band circularly-polarized square microstrip antenna," *IEEE Trans. Antennas Propag.*, Vol. 49, No. 3, pp. 377–382, 2001.

[19] A. Vallecchi, J. R. D. Luis, F. D. Flaviis, "Low profile fully planar folded dipole antenna on a high impedance surface," *IEEE Trans. Antennas Propag.*, Vol. 60, No. 1, pp. 51–62, 2012.

[20] Y. Dong, H. Toyao, T. Itoh, "Compact circularly-polarized patch antenna loaded with metamaterial structures," *IEEE Trans. Antennas Propag.*, Vol. 59, No.11, pp. 4329–4333, 2011.

[21] S. X. Ta, I. Park, R. W. Ziolkowski, "Circularly polarized crossed dipole on an HIS for 2.4/5.2/5.8-GHz WLAN application," *IEEE Antennas Wireless Propag. Lett.*, Vol. 12, pp. 1464–1467, 2013.

[22] H. Mosallaei, K. Sarabandi. "Antenna miniaturization and bandwidth enhancement using a reactive impedance substrate," *IEEE Trans. Antennas Propag.*, Vol. 52, No. 9, pp. 2403–2414, 2004.

[23] K. Agarwal, Nasimuddin, A. Alphones, "RIS-based Compact Circularly Polarized Microstrip Antennas," *IEEE Trans. Antennas Propag.*, Vol. 61, No.2, pp. 547–554, 2013.

[24] L. Bernard, G. Chertier, R. Sauleau, "Wideband circularly polarized patch antennas on reactive impedance substrates," *IEEE Antennas Wireless Propag. Lett.*, Vol. 10, pp. 1015–1018, 2011.

[25] K. Agarwal, Nasimuddin, A. Alphones, "Triple-band compact circularly polarized stacked microstrip antenna over reactive impedance meta-surface for GPS applications," *IET Microw. Antennas Propag.*, Vol. 8, No. 13, pp. 1057–1065, 2014.

[26] K. Agarwal, Nasimuddin, A. Alphones, "Wideband circularly polarized AMC reflector backed aperture antenna," *IEEE Trans. Antennas Propag.*, Vol. 61, No. 3, pp. 1455–1461, 2013.

[27] K. Agarwal, Nasimuddin, A. Alphones, "Design of compact circularly polarized microstrip antennas using meta-surfaces," *43rd European Microwave Conference* (EuMC'2013), Nuremberg, pp. 1067–1070.

[28] H.-X. Xu, G.-M. Wang, M. Q. Qi, "Compact dual-band circular polarizer using twisted Hilbert-shaped chiral metamaterial," *Opt. Express*, Vol. 21, No. 21, pp. 24912–24921.

[29] T. Sato, K. Shiraishi, K. Tsuchida, et al. "Laminated polarization splitter with a large split angle," *Appl. Phys. Lett.*, Vol. 61, pp. 2633–2634, 1992.

[30] T. Cai, G.-M. Wang, F.-X. Zhang, J.-P. Shi. "Low-profile compact circularly-polarized antenna based on fractal metasurface and fractal resonator," *IEEE Antennas Wireless Propag.* Lett., Vol. 14, pp. 1072–1076, 2015.

[31] J. Sun, J. Li. "Terahertz wave polarization splitter using full band-gap photonic crystals," *J. Infrared Millim. Terahertz Waves*, Vol. 36, No. 3, pp. 255–261, 2015.

[32] S. Harish, X. Xu, H. Amir, et al. "Recent advances in silicon-based passive and active optical interconnects," *Opt. Express*, Vol. 23, No. 3, pp. 2487–2510, 2015.

[33] J. Hyung Lee, J. W. Yoon, M. J. Jung, et al. "A semiconductor metasurface with multiple functionalities: A polarizing beam splitter with simultaneous focusing ability," *Appl. Phys. Lett.*, Vol. 104, pp. 233505, 2014.

[34] Y. Xu, J. Xiao, X. Sun. "Proposal for compact polarization splitter using asymmetrical three-guide directional coupler," *IEEE Photon. Technol. Lett.*, Vol. 27, No. 6, pp. 654–657, 2015.

[35] J. Zhao, Y. Chen, Y. Feng. "Polarization beam splitting through an anisotropic metamaterial slab realized by a layered metal-dielectric structure," *Appl. Phys. Lett.*, Vol. 92, pp. 071117, 2008.

[36] H. Luo, Z. Ren, W. Shu, et al. "Construct a polarizing beam splitter by an anisotropic metamaterial slab," *Appl. Phys. B*, Vol. 87, pp. 283–287, 2007.

[37] H. F. Ma, G. Z. Wang, W. X. Jiang, et al. "Independent control of differently-polarized waves using anisotropic gradient-index metamaterials," *Sci. Rep.*, Vol. 4, pp. 6337, 2014.

[38] T. Cai, G.-M. Wang, F.-X. Zhang, et al. "Ultra-thin polarization beam splitter using 2D transmissive phase gradient metasurface," *IEEE Trans. Antennas Propag.*, Vol. 63, No. 12, pp. 5629–5636, 2015.

[39] N. Yu, P. Genevet, M. A. Kats, et al. "Light propagation with phase discontinuities: generalized laws of reflection and refraction," *Science*, Vol. 334, pp. 333–338, 2011.

[40] C. Pfeiffer and A. Grbic. "Cascaded metasurfaces for complete phase and polarization control," *Appl. Phys. Lett.*, Vol. 102, pp. 231116, 2013.

[41] J. R. Cheng and H. Mosallaei. "Optical metasurfaces for beam scanning in space," *Opt. Lett.*, Vol. 39, No. 9, pp. 2719–2721, 2014.

[42] J. Luo, H. Yu, M. Song, et al. "Highly efficient wavefront manipulation in terahertz based on plasmonic gradient metasurfaces," *Opt. Lett.*, Vol. 39, No. 8, pp. 2229–2231, 2014.

[43] Z. Wei, Y. Cao, X. Su, et al. "Highly efficient beam steering with a transparent metasurface," *Opt. Lett.*, Vol. 21, No. 9, pp. 10739–10745, 2013.

[44] Y. F. Li, J. Q. Zhang, S. B. Qu, et al. "Wideband radar cross section reduction using two-dimensional phase gradient metasurfaces," *Appl. Phys. Lett.*, Vol. 104, pp. 221110, 2014.

Permissions

All chapters in this book were first published in MAS, by InTech Open; hereby published with permission under the Creative Commons Attribution License or equivalent. Every chapter published in this book has been scrutinized by our experts. Their significance has been extensively debated. The topics covered herein carry significant findings which will fuel the growth of the discipline. They may even be implemented as practical applications or may be referred to as a beginning point for another development.

The contributors of this book come from diverse backgrounds, making this book a truly international effort. This book will bring forth new frontiers with its revolutionizing research information and detailed analysis of the nascent developments around the world.

We would like to thank all the contributing authors for lending their expertise to make the book truly unique. They have played a crucial role in the development of this book. Without their invaluable contributions this book wouldn't have been possible. They have made vital efforts to compile up to date information on the varied aspects of this subject to make this book a valuable addition to the collection of many professionals and students.

This book was conceptualized with the vision of imparting up-to-date information and advanced data in this field. To ensure the same, a matchless editorial board was set up. Every individual on the board went through rigorous rounds of assessment to prove their worth. After which they invested a large part of their time researching and compiling the most relevant data for our readers.

The editorial board has been involved in producing this book since its inception. They have spent rigorous hours researching and exploring the diverse topics which have resulted in the successful publishing of this book. They have passed on their knowledge of decades through this book. To expedite this challenging task, the publisher supported the team at every step. A small team of assistant editors was also appointed to further simplify the editing procedure and attain best results for the readers.

Apart from the editorial board, the designing team has also invested a significant amount of their time in understanding the subject and creating the most relevant covers. They scrutinized every image to scout for the most suitable representation of the subject and create an appropriate cover for the book.

The publishing team has been an ardent support to the editorial, designing and production team. Their endless efforts to recruit the best for this project, has resulted in the accomplishment of this book. They are a veteran in the field of academics and their pool of knowledge is as vast as their experience in printing. Their expertise and guidance has proved useful at every step. Their uncompromising quality standards have made this book an exceptional effort. Their encouragement from time to time has been an inspiration for everyone.

The publisher and the editorial board hope that this book will prove to be a valuable piece of knowledge for researchers, students, practitioners and scholars across the globe.

List of Contributors

Mohammad Alibakhshikenari and Ernesto Limiti
Department of Electronic Engineering, University of Rome Tor Vergata, Rome, Italy

Mohammad Naser-Moghadasi
Faculty of Engineering, Science and Research Branch, Islamic Azad University, Tehran, Iran

Ramazan Ali Sadeghzadeh
Faculty of Electrical Engineering, K. N. Toosi University of Technology, Tehran, Iran

Bal Singh Virdee
Center for Communications Technology, London Metropolitan University, London, UK

Mayumi Matsunaga
Department of Electrical and Electronic Engineering, Ehime University, Matsuyama, EhimeJapan
Research Institute for Sustainable Humanosphere, Kyoto University, Uji, Kyoto, Japan

Bin Zhou, Junping Geng, Xianling Liang, Ronghong Jin and Guanshen Chenhu
Department of Electric Engineering, Shanghai Jiao Tong University, Shanghai, China

Ridhwan Khalid Mirza and Yan (Rockee) Zhang
Intelligent Aerospace Radar Team, Advanced Radar Research Center, School of ECE, University of Oklahoma, Norman, USA

Richard Doviak and Dusan Zrnic
National Severe Storm Laboratory, NOAA, Oklahoma, Norman, USA

Keyhan Hosseini and Zahra Atlasbaf
Faculty of Electrical and Computer Engineering, Tarbiat Modares University, Tehran, Iran

Paras Chawla
Electronics and Communication Engineering Department, Chandigarh Engineering College Landran, Greater Mohali, Punjab, India

Rohit Anand
Electronics & Communication Engineering Department, G. B. Pant Government Engineering College, New Delhi, India

Sudipta Chattopadhyay
Department of Electronics and Communication Engineering, Mizoram University, Aizawl, Mizoram, India

Subhradeep Chakraborty
TWT Group, MWT Division, CSIR-Central Electronics Engineering Research Institute, Pilani, Rajasthan, India

Mohammad Alibakhshikenari and Ernesto Limiti
Department of Electronic Engineering, University of Rome Tor Vergata, Rome, Italy

Tong Cai, He-Xiu Xu, Guang-Ming Wang and Jian-Gang Liang
Microwave Laboratory, Air Force Engineering University, Xi'an, China

Index

A

Anechoic Chambers, 50-52, 65, 68
Ant Colony Optimization, 99, 102-103, 115-116
Antenna Coordinate System, 52
Antenna Structures, 2-3, 7-9, 18, 97, 140, 151
Artificial Bee Colony Optimization, 100, 102-103, 115

B

Bat Algorithm, 100, 102-103, 115
Beam Scanning, 83-84, 92-95, 194
Beam Steering, 82, 87, 194
Bias Distribution, 75-76
Biogeography-based Optimization, 100, 102-103, 117

C

Circularly Polarized Antennas, 15
Coaxial Cable, 15, 23-27
Coaxial Cylinder, 30-35, 39-40, 45, 47-48
Constant Current Loop, 55
Coplanar Waveguide, 84, 146, 155-156
Cross Polarization, 6, 50-51, 54, 67-68, 70, 144
Crosseddipole, 16-19
Cross shaped Loop Antenna, Cuckoo Search, 99, 102-103, 116

D

Dielectric Resonator Antennas, 30, 48
Differential Evolution, 101-103, 115-116
Dipole Antennas, 18, 29, 58
Dipole Feeder, 25-27, 29
Dispersion Diagram, 9, 72, 74-75, 81, 83, 87-89, 93, 148
Dual Polarization Elements, 51

E

E&m Dipole Array, 50, 65
Electric Dipole, 50-51, 58-63, 65, 68

F

Fabricated Prototypes, 2-3
Feeding Mechanism, 15, 24-25
Feeding Port, 19, 22-23, 25, 35
Ferrite Loaded Waveguide, 82
Firefly Algorithm, 99, 102-103, 115
Fractal Resonator, 162, 172, 175, 193

G

Galaxy-based Search Algorithm, 100, 115
Genetic Algorithm, 99, 102-103, 116
Ground Planes, 2, 7, 18

H

Harmony Search, 100, 102-103, 115-117
Heterodyne Mixing, 79, 94

I

Impedance Matching, 3, 32-33, 57, 169, 175, 177, 179, 186, 192

L

Leaky Wave Antennas, 72, 94
Lindenblad Antenna, 17-18
Linear Polarization, 67-68, 90
Loop Antennas, 15, 18, 55-56, 70

M

Magnetic Conductor, 82, 173
Magnetic Dipole, 40, 50-51, 55-56, 58, 61, 63, 68
Metamaterials, 2, 13-14, 72, 94-95, 145, 161-162, 180, 191-192, 194
Microstrip Antennas, 96, 118-119, 140-143, 161, 191, 193
Microstrip Patch Antennas, 18, 86, 95, 97, 118, 141-142, 162
Miniaturized Antennas, 1
Monopole Radiator, 1-2
Monopoles, 2, 10, 31
Multiband Antennas, 1, 158
Multipolarization, 18, 27
Multi function Phased Array Radar, N Network Analyzer, 3, 109, 169, 179, 186
Noncellular Communication, 1

O

Omnidirectional Circularly Polarized Antenna, 30, 48-49, 192
Omnidirectional Plane, 31, 37-39, 43-46
Operation Bandwidth, 43, 45

P

Particle Swarm Optimization, 48, 99, 102-103, 115-116
Phase Shifter, 16
Planar Antennas, 1, 14, 118, 145

R

Radiating Element, 50-51, 61, 68
Radiofrequency Identifier, 15-16

Reconfigurable Antenna, 96-98, 100, 109-110, 112-114, 117
Reflecto-directive System, 79-80, 94
Resonant Frequency, 2, 4, 25-26, 118-130, 136-137, 142-143, 166, 168-169, 174-175

S

Simulated Annealing, 101-103, 116
Slot Array, 30-34, 39-40, 47, 49
Soil Scattering Information, 51
Spiral Antennas, 15, 29
Split-ring Resonator, 1, 3
Stub Matching, 84
Substrate Integrated Waveguide, 72, 76, 80, 94-95
Substrate Permittivity, 3, 131

T

Transition Frequency, 74, 77, 79, 83, 86, 90-91
Transmission Line, 13-14, 29, 33, 49, 72-73, 75-76, 94, 98, 104, 107, 109, 142, 145-146, 155-156, 160, 192
Tunable Radiation Angle, 74, 94
Turnstile Antenna, 16, 28

U

Ultra-wideband, 1

V

Varactor Diodes, 75-76

W

Weather Surveillance, 50
Wireless Communication, 1, 14, 30, 140, 145, 161-162, 172, 180, 188
Wireless Power Transmission, 16
Wireless Technologies, 96, 113, 159

www.ingramcontent.com/pod-product-compliance
Lightning Source LLC
Chambersburg PA
CBHW080258230326
41458CB00097B/5109
* 9 7 8 1 6 3 9 8 7 1 8 9 6 *